First Published - 2008

Reprinted - 2016

ISBN: 978-81-8356-321-5

Environmental Pollution and Hazards

Published by:

DISCOVERY PUBLISHING HOUSE PVT. LTD.
4383/4B, Ansari Road, Darya Ganj
New Delhi-110 002 (India)
Phone: +91-11-23279245, 43596064-65
Fax: +91-11-23253475
E-mail: discoverypublishinghouse@gmail.com
sales@discoverypublishinggroup.com
web: www.discoverypublishinggroup.com

Printed at:
Infinity Imaging Systems
Delhi

Preface

The surroundings we live in is environment which become unhealthy in the pace of time arrived in 21st century i.e. the modern era of industrial revolution, full of radio waves in the air, unhygienic gases an air and chemical substances on earth and water, poisonous elements in food and breath, depletion of forest, soil erosion, bio-diversity, extinction of tress and plants of medicinal value, rapid construction of domestic and commercial houses and buildings, heavy dust and humidity, irregular rainfall, increasing of heat on air, green house effect, holes on ozone layer, less oxygen in air, ground water decreasing, increasing of dry land, water crisis, etc. are the alarming factors to survive the mankind on the earth. Day-by-day these problems arises in a furious mode. To check these problems men should be conscious about all the problems and remedial measures to be taken by them universally. So check it, plantation and aforestation, checking of deforestation, water shed management, rain water-harvesting, less use of chemicals industrialisation with environmental safeguards, interlinking of big rivers, sanitation facilities in both rural and urban areas, garbage management and recycling, preservation of medicinal plants, protection of flora and fauna, etc. are the prime concerns of environmental conservations which should follow environmental audit and innovations should continue on it. The paper contributors have been initiated their research and experiments at their level best to make the book sound and the publisher extended their helping hands to release it.

Contents

1

Culture Based Traditional Environmental Knowledge and Role of Rural Women in Environment Conservation

A Case Study

Miss Supriya Prusty♣
Dr. Jagadish Kumar Kar♠

Introduction

Earth is the only planet so far known with an environment, necessary for sustainable life. In this environment human beings have originated and survived since time immemorial. Land, water, air, plants and animals are the main components of the environment. Survival of human beings and other living creatures at present and in the period ahead is largely depending upon natural equilibrium amongst all these elements. Normally, we want to breathe fresh air, use pure water, and live in noiseless atmosphere. But unfortunately, we find ourselves surrounded by dangerously polluted air, water and unbearable noise. All these combined together may be called environmental

♣ Lecturer, Department of Sociology, Kshetrabasi DAV College, Niakarpur, Dist-Khurda, Orissa, Pin-752 010.

♠ Lecturer, Department of Zoology, Kshetrabasi D A V College, Niakarpur, Dist-Khurda, Orissa, Pin-752 010.

pollution, which have emerged as a major problem against peace and security of human existence. At present, development, environment and peace are the issues of significant importance for a nation and the world as a whole. These three have an interesting interrelationship among each other, which is debated at the time of crisis. The concern of academics, environmental experts and scientists for these issues, arise from their painful awareness of the reality that all kind of development are attended by the specter of natural environment and rapid depletion of natural resources. As a result, development creates the theoretical illusion of progress, mental and material satisfaction and peace; but on the other hand, all its variations, becomes a casualty.

In fact, environmental pollution is not a new problem. When man and other higher animal began their life on this earth, there was absolutely no sign of pollution. There was perfect balance between various natural processes. The air and water were pure and the soil was fertile. The problem of pollution arose with the dawn of the civilisation. As soon as man learned to use fire, the pollution began. As human population increased, there was increase in our working sphere and with this also increased pollution. Yet civilisation still depends on the natural environment not only for energy but also for vital life support processes such as air and water. The basic law of nature has not been repealed; but, only their complexion and quantitative relations have changed, as the world's human population and its prodigious consumption of energy have increased our power to alter our environment. Accordingly, our survival depends on knowledge and intelligent action to preserve and enhance environmental quality by means of harmonious rather than disruptive technology.

Environment and Society

The interconnection between human beings and their environment is a complex one. The environmental problems are created through exploitation of nature and natural environment by human beings. Thus the problems of "taking too much from the environment and restoring too little" need an intelligent and helpful solution.

Nature has been a part of Indians' social life. It is found that, trees, plants, streams, waterfalls, rivers, lakes, seas, hills, mountains, snow, rain , sky, etc. have been intertwined with every kind of human experience and seriously caught the attention of poets, philosophers, nature lovers, social activists and environmentalists. Today the average Indian citizen is still very much dependent on his natural environment for the satisfaction of his/her basic needs. As an occupation agriculture is very much intertwined with nature. Factors like, water, soil and atmosphere greatly shape the kind and extent of agricultural activities in India.

Nature is accepted as divine in Hinduism. Hinduism expresses firm belief that the natural environment in which people are placed is a manifestation of divine nature. Various trees and plants are found to be used in religious ceremonies and worship. Among them some trees are considered so sacred that particular Gods and Goddesses have made their abode in them. The *Vedas, Puranas, Upanishad* and other scriptures of Hindu religion give a detailed description of trees, plants and wild lives, and their importance to the community. In *Mahabharata* it is said that in a village even if there is only one tree full of flowers and fruits that become worthy of worship and respect. Even *Varaha Purana* advocates regular plantation as a means to achieve heaven.

Felling green trees, klilling birds and animals are strictly prohibited in ancient Hindu scriptures. While in *Yajur Veda* it is said that no person should kill animals, *Artha Shastra* explains how those people should be punished who dirt the public places. In *Bhagavad Gita,* Krishna explicitly mention that he is the creator of everything in the world including man and, therefore, he has to have the attitude of *'Sarvabhuta Hiteratah'* that is taking delight in all other beings in the world.

The foundation for environmental protection in the Islamic faith is based on the religious principle that Allah/

God created all aspects of the natural world with different functions that were determined by Allah/God. The tribal culture indicates that the earth is a living conscious being that must be treated with respect and loving care. Environment and natural resources are God's gifts to man and other living beings. Man has a "moral responsibility" to share them with others and also to keep them safe for the future. Philosophers have commented that "the world is beautiful but has a disease called man". Instead of being a "liability" man should be an "asset" to the world. (Mishra, 2005, p. 45).

Environment and Social System

There are symbiotic relationships between bio-physical ecosystems and social systems, with strong cultural interconnections between the two. Traditional societies have co-evolved with their environment, modifying nature but actively maintaining it in a diverse and productive state, based on locally evolved traditional ecological knowledge; socio-cultural practices and/or religious beliefs since antiquity. In Indian culture-linked environmental management is found in many of the mountain societies, who have many natural resource linked institutions. These sacred institutions rather than promoting environmental consciousness per se intended to boost religious norms by fostering social solidarity and conservation values. It is in the context of ensuring community participation for effective management of natural resources on an integrated approach with culture-based Traditional Ecological Knowledge (TEK) as the interconnection element between ecological, socio-cultural and economic dimension is required (Ramakrishnan, 2001). Many traditional societies all over the world value a large number of plant species from the wild and sometimes animal species too, for both economic and/or spiritual reasons.

Hindu culture believes in the partnership and stewardship ethics, which stands for holding of land (including trees, rivers and mountains, and animals) in trust for God and general benefit of mankind. Man is the trustee of the universe and he is authorised to use natural resources but has no divine power of control and dominion of nature and its elements. Hence, from the perspective of Hindu culture, abuse and exploitation of the nature for immediate gain is unjust and irreligious (Ignacimuthu, 1998).

Environment and Traditional Medicine

The traditional system of medicine, *Ayurveda* in India, for example, has many medicinal formulations from *Neem* and Tulsi trees. The recent commercial patenting of the active principle from these trees have been successfully extracted chemically and patented for their pesticides properties. It shows that attempts have been made to exploit the natural resources. There are also a variety of traditional medical practices including the *Unani* and *Siddha* systems restricted in northern and southern part of India respectively.

Traditional medicinal plants are the priceless gift of nature and Indians since time immemorial have been depending on herbal medicine. Rural people till today depend upon nature for their health care system. It is mandatory for every Hindu family to grow *'Tulsi van'* within the boundary of the house and one or two plants especially to be available for daily worship. A simple Hindu rural woman spends considerable time around this plant doing worship and lighting lamps. Rural women have applied their own indigenous knowledge for self-cure and their family members. Some common herbs and shrubs like, *Tulsi (Ocimum sancturm* Linn.), Turmeric *(Curcuma longa* Linn.), *Adosa (Adhatoda vasica* Needs), *Sant (Borehavia diffsa* Linn.), *Pipali (Piper longum* Linn.), Ginger *(Zingiber officinale* Roseoe), *Brahmi (Bucopa mooniera* Linn.), *Til (Sesamum idicumum* Linn.), *Giloe (Tinospora cordifolia* Miers), *Salsa (Hemidesmus indicus R.* Br.) etc. are used for medicinal value in our day to day life. Role of these plants in the rural health care system is of crucial importance. For example, *Tulsi* is also a good medicine for common fever, cough and coryza among children. The daily use of Turmeric powder internally and topically improves flesh tone and colour of the skin. *Adosa* is expectorant, antispasmodic, a blood purifier and a tonic. It is also laxative, diaphoretic and anti-helminthic. Pipali is carminative, cardiac stimulant, digestive, diuretic emollient, an appetiser, a vermifuge, and rejuvenator. Ginger is an active ingredient of various preparations used for dyspepsia and piles. *Brahmi* is administered as a nervine, cardiac tonic and as febrifuge. It is very useful in nervine disorders and mental diseases. (Mani et al. 1997, pp. 29-32.).

Environmental Degradation: A Serious Global Concern

Environmental degradation has become a serious problem all over world. Air and water pollution combined with mismanagement of solid wastes have caused serious damage to the equality of the human life. The other disaster include change in fertility and mortality, loss of comprehensiveness, reduced output in resource based sectors, ecological imbalance, loss of bio-diversity, acid rain greenhouse effect, global warming and natural calamities like, tsunami, cyclone, flood, draught etc. The most important cause however, is growth of population and resultant rise in poverty. Besides, economic development, urbanisation, industrialisation, availability of natural resources and the life style of its people also affect the environment.

Effects and Causes of Environmental Degradation

Environmental problems have attracted the attention of a wide cross-section of people all over the world during the last three decades. India which is already engaged in bring population above poverty line is put to still greater pressure towards environmental protection. Poverty is the fundamental cause for over exploitation of natural resources of the country. Mrs. Indira Gandhi (1972) addressed to the Stockholm Conference that "Poverty and need are indeed the greater polluters"

Today the cry of "pollution" is heard from all nooks and corners of the world, and pollution has become a major threat to the existence of mankind on this earth. The pollution of various sources has gone to such an extent that we are unable to breathe fresh air, drink fresh water and sleep peacefully due to different type of psychosomatic disorders. The factors responsible for environmental pollution are increased population pressure, urbanisation, industrialisation, automation, unsound planning and lack of general awareness. Migration from rural to urban set up has resulted in the problems like, congestion, insanitation, slums and degradation of environment. In most of our cities congestion and overcrowding create problems to public health and life styles of the people. The setting up of industries in residential areas or areas near to urban dwellings, and the delivery of industrial discharges directly to air and water sources without any preventive measures seem to contaminate the environment with toxic and inorganic contaminants. Production and use of chemical fertilisers and pesticides for agriculture and horticulture affect both air and water severely. The atomic and thermo-nuclear explosions both in air or underground or underwater are having a great impact on environment. Similarly, noise from industries, vehicles, street noise (loud speaker, marriage procession, demonstration, etc.) and mass media (radio, tape recorder and TV) have considerable negative impact on environment. The restless life, strained nerves and the tensed mind are the consequences of noise pollution in our daily lives.

Global Issues on Environment

In the global context, the issues like, global warming and green house effect, the CFCs (Chlorofluorocarbons) problem and the Ozone hole controversy, acid rain, deforestation, nuclear hazards and high level of radioactivity detected in vegetation, milk, oil, food, water, air and soil are some of the major challenges on the face of mankind.

Green house effect and climate change are the global phenomena. Carbon dioxide is the most abundant and important greenhouse gas (gas which causes warming). The present level of carbon dioxide in the atmosphere is of the order of 350 ppm (parts per million). The green house warming due to this level of carbon dioxide keeps the atmosphere of the earth warm enough to sustain life. If the present rate of increase of greenhouse gas levels in the atmosphere remain unabated (doubling of CO_2 level to 550 ppm is possible in the next 50 years or so), the rise in the global atmospheric temperature can result in a change in the global climate with profound and far reaching consequences. Ozone is a naturally occurring gas in the atmosphere produced by virtue of atmospheric reactions. The ozone layer protects the living organisms on the earth surface from the hazards of intense ultraviolate light from the sun and cosmic radioactive rays. Greater exposures to ultraviolate light and cosmic radiations due to ozone depletion increase the chance of skin cancer and weaken the body immune system. This has been a matter of great concern for the international community.

Deserts are now being expanding in various parts of the world; as per report an area equivalent to about nine-baseball field is turned in desert every minute (Bandapadhyay, 1993, pp. 14-15). Dying oceans, declining whale and tiger population, radioactive fall out from nuclear test and many more clearly indicate man's uncontrolled greed and indifferences over nature to the need for preserving his fellow creatures. Projection of an exploding world population raised the fears that the world would run short of food and resources and the last vestiges of wilderness would be on the verge of extinction. The first United Nations Conference on "Human Environment" held in Stockholm in 1972 was the culmination of gathering environmental concern of the 1960s. But, after three decades of this great conference environment degradation today is more acute. World population has increased nearly about 5 billion (20%) with more mouths to be feed and a lot more provided a clear environment and better quality of life. The pressure on the Governments, both in the developing and developed worlds is mounting toward higher priority to environmental issues and seeks urgent action on them.

Naturally, environment has become a global issue with many ramifications. Better environment management for providing safe and portable drinking water, to check global warming and to maintain biodiversity is one of the hot topics for discussion today at national, regional and global levels. All these put in toto the big question who is the cause of this polluted environment.

Environmental Degradation in India

In India, the greatest pollution centres are seen in industrial townships. As per the report, Delhi, Kolkata and Mumbai have occupied 4th, 8th, 13th ranks among the highly polluted towns of the world (The Samaj, 2nd June, 1995). The industries discharge their wastes directly into the environment as gaseous wastes (smokes, sulphur dioxide, carbon monoxide and other poisonous gases); solid wastes (dust particles and other inorganic and organic solids) and liquid wastes (effluents) are making the environment being polluted. Residents of Delhi are today exposed to highest levels of air pollution in the country and perhaps the world. The emission of gases like sulphur dioxide and oxides of nitrogen from Mathura based petroleum refinery is poisoning threat to Tajmahal at Agra and other monuments at Fatepursikri. Thermal power station at Singrauli in UP, Korba in M.P., Farakka in West Bengal and Talcher in Orissa releasing fly ash, sulphur dioxide and various hydrocarbons, and polluting the air and also spreading the serious respiratory diseases like asthma, bronchitis and lung cancer etc. Plants also suffer a lot by these pollutants. Similarly, by receiving the effluents of the major industries along the banks, the river Ganga, Cauvery, Jamuna, Mahanadi and Rushikulya are being polluted. The vehicular noises of the busy cities like Delhi, Bombay, Kolkata and Bhubaneswar, the capital of Orissa also creating noise pollution as well as their exhaust increasing the carbon dioxide and sulphur dioxide contains in the air. Thus, major cities and towns in India are polluted by the impact of industrial development and lack of civic sense of knowledge on hygiene and sanitation.

Environmental Degradation Scenario in Orissa

In Orissa, noise pollution is restricted to the busy town like Cuttack, Bhubaneswar, Puri, Rourkela, etc. These are places of great importance. As Cuttack, is the ex-State capital and business centre, Bhubaneswar is the present State capital, Puri is the world's most emphasised tourist spot and Rourkela is the area of most industrial importance. The air of these towns is polluted by the smoke emissions of the nearby industries and automobile exhausts. The environment of some other places like Chowdwar, Rayagada and Titilagarh is some extent polluted by the smoke as well as paper mill effluents discharged to the environment. The Cement industries at Rajgangpur and Baragarh are also creating pollution of local environment. The environment of Angul, Talche and Kaniha is being polluted due to NALCO plants, Thermal Power-Plant, Fertiliser complex and Colliery activities. Eco-degradation is seen in Joda, Barbil and Bansapani due to excessive mining and quarrying. Environmental pollution is also noticed at Balasore due to industries like, Birla tyres, Polar latex, Emami Papermills, Ispat Alloys and other medium and small industries. Similarly, Paradeep is being polluted by Aswal fertilisers plant (recently taken over by IFFCO—Indian Farmers Fertiliser Company) and Paradeep Phosphate Limited (PPL Ltd.). Several other industries and factories located in various places seem to be the contributor of environmental pollution and ecological degradation in Orissa. Some of these are: Ferro-Alloys Corporation and Charge Chrome Ltd. at Bhadrak; Indian Charge Chrome Ltd. at Chowdwar; MESCO Steels and Neelanchal Ispat at Duburi in Jajpur district; Arati Steels at Ghantikhal in Cuttack district and many more. The effluents of Jayashree Chemicals near Bhanjanagar in Ganjam district are polluting the river Rushikulya. The issues like environmental degradation and eutrophication of Chilika Lake by prawn farming seem to decreasing the nesting grounds of inland and migratory birds. Eco-degradation in Bhitarakanika is also a burning issue in Orissa, which is detrimental to sea Trustless (Olive Ridley) and Crocodiles.

The seashore at Puri, Konark and Paradeep, are spoilt by cutting trees abruptly, building new five star hotels restaurants and shops etc. as the sign of civilisation and development.

Orissa has also signed MoU (Memorandum of Understanding) with an array of Industrial Groups (nearly 38 numbers) for the establishment Steel Plant and mining of mineral ores. Environmentalists, local people even opposition political leaders are seem to protesting against establishment and mining because of ecological degradation, destruction of forest cover and demanding resettlement of the inhabitants of the plant and mining covering areas, as a result to these Lanjigad and Kalinganagar had occupied in news headlines in the recent past. No doubt, establishment of these plants shall definitely bring about industrial and economic development, on the other hand, the quantity of pollutants will be more and the quantum of pollution will also very high in the future. This will definitely affect the citizens of Orissa if necessary attention is not given for pollution control, affectation, soil conservation and management of environment and natural resources etc. In Orissa rural areas also not free from pollution. The rural areas around the industrial establishments, cities and towns are facing the problems of environmental pollution due to smokes, ash, effluents etc. Besides these, maximum Crusher industries and rice mills are found located in rural areas. In rural areas pollution due to open defecation, urination, burning of dead bodies, washing and bathing of animals are very common. All these seem to create environmental pollution and ecological degradation in several parts of rural Orissa.

Orissa has also experienced nature's fury time and again situated on the coast of Bay of Bengal; the State is extremely vulnerable to multiple hazards, most of which are water/ wind related. In 1998, more than two thousand people fell to the severest heat wave to the last century. In 1999, nearly nine thousand human lives were lost and millions were rendered homeless in the Super-cyclone, one of the gravest calamities of the millennium. The year 2003 has so far witnessed two calamities. The state reeled under oppressive heat during the summer, at Titilargarh mercury soared up to 50.1 degree Celsius, which was the highest far any place of in the State. The highest temperature recorded at Bhubaneswar was 44.9 degree Celsius (Mohanty, 2003, pp. 74-76).

Heavy rainfall triggered off the second calamity of the year. Incessant downpour in the upper and lower catchments of river Mahanadi in the last week of August spelt disaster. Heavy rains in Kalahandi and Bolangir districts caused floods in the tributaries Hati and Tel, which marked the beginning of crisis. Rains in Chhatishgarh worsened the situation, pushing the water level in Harakud dam close to the full reservoir level of 630 feet. Flood came in four successive waves, affecting different parts of the State namely, Bolangir, Kalahandi, Sonepur and several other parts of the Western Orissa. In 2006, Orissa had witnessed the severe flood in the catchments of rivers in Koraput, Raygada, Sonepur, Mayurbhanj, Balasore, Bhadrak, Jajpur, Cuttack, Kendrapara and Jagatsinghpur districts.

Environmental Degradation in Tangi Block

Tangi Block with its geographic, climatic and biological diversity has a unique ecological environment. In its territorial area deforestation, soil erosion above all environmental pollution is most frequent. This area has also witnessed the super cyclone in 1999. So far climatic condition is concerned, the summer is very hot and dry, rain is humid with normal to heavy rainfall and winter is chill with low humidity content in the environment. The other seasons are occasionally felt with seasonal climates. This area not only experiences natural disasters but also pollution due to crusher industries, rice mills, automobile exhausts and several day to day anthropogenic activities. The contributing factors of air pollution of this locality are stone dusts from crusher industries, vehicular exhausts from the motor cars running on the National Highway No—5 running between Howrah and Chennai. Besides these, exhausts from railway engines running on East-coast railway line passing through the area, throwing domestic garbage in the open, defecation in the open, smokes from household *chullas* (cooking stove) and burning of dead bodies, dusts from agriculture wastes and post harvesting practices, harmful chemical particles from pesticide and insecticide spray are some of the compelling causes of air pollution. Similarly, the contributing causes of water pollution are effluents from rice mills; sewage from household kitchens, hotels and restaurants, run off water from the chemical fertiliser and pesticide applied crop fields etc. The immersion of coloured and painted earthen statue along with flowers in to the water bodies though reflects holy culture of Orissa still contribute to pollute the water bodies. Ground water pollution through increased fluoride content and surface water contamination is also reported (The Sambad, 2005, p. 10). Agriculture wastes from the post harvesting practices are also contributed for air and water pollution. The major sources of noise pollution are vehicular noise from trains and motorcars and high sound music of the celebrations of several social functions. As a consequence to environmental pollution filarial is found endemic in this area besides, various skin diseases and gastro intestinal problems are found frequently in the population. Socio-cultural degradation is not prominently marked, however younger generations are found marching towards modernisation in their dress and living styles. The cultural practices in relation to environment are also found loosing its importance.

Environment Preservation and Conservation Measures at the Global Level

The international effort for the protection and preservation of environment became vigorous in the late 1970s and 1980s.

1. The United Nations' (U.N.s) Conference of Human Environment held in Stockholm, Sweden 1972. This was the first international conference, especially dealing with the global environmental concerns.

2. The U.N.'s Habitat Conference on Human Settlements in Vancourver, Canada in 1976.
3. The World Water Conference in Mardel Platu, Agrentina in 1977.
4. The U.N.'s Desertification Conference held in Nairobi in 1977.
5. The United Nations Environmental Programme (UNEP) and United Nation Educational, Scientific and Cultural Organisation (UNESCO) organised an Intergovernmental Education Conference at Georgia in 1977.
6. World Commission on Environment and Development (WCED) or Brundtland Commission (1984) set up in pursuance to the U.N. General Assembly Resolution in 1983 to re-examine the critical environment issues and to formulate proposals for them.
7. The Paris Conference, held in 1986, was called for saving trees and forests.
8. The U.N. convened a United Nations' Conference on Environment and Development (UNCED) at Rio de Janeiro in 1992. It is called the Earth Summit. A comprehensive document, called Agenda -21 was issued at this conference.

   The Rio Declaration spelt out general principles for sustain able development. The Agenda-21 document covered a wide variety of conservation issues and management of resources for development so that the future needs for resources were taken care of The atmosphere, land resources, forests, mountains sustainable agriculture and rural development, drought, biological diversity, seas and ocean, fresh water, and waste management were among the subjects that were covered with intensity with a programmatic content that made the humanity for a better 21st century in a sustainable mode.

9. The World Summit on Sustainable Development was held at Johannesburg (South Africa) in August-September, 2002. It generally reviewed the progress of implementation Agenda-21 of the Rio Summit Declaration of 1992, in order to satisfy whether the aims and objectives set forth before the Nations and the UN and its Agencies have been by and large achieved.
10. The Eighth conference under the UN Framework Convention on Climate Change was held at New Delhi (India) in October-November, 2002. It linked Sustainable Development with climate change.

Environment Protection Measures in India: Constitutional Provisions

The term 'environment' does not find place as a subject on the Union, State or Concurrent List on 7th Schedule of the Indian constitution. As residuary subject, the Union Government has powers to enact laws on environment. However, in the 42nd amendment of the constitution two articles having direct bearing on environment were inserted in 1976. These are:

Article 48 A—The state shall endeavour to protect and improve the environment and to safeguard the forest and wildlife of country. It is imposing a duty on the state to protect environment, forest and wildlife of the country.

Article 51 A (G)—It is the duty of the citizen to protect and improved the natural environment including forests, lakes and rivers and wildlife and to have compassion for all living creatures.

Besides these, ***Article 21*** of the constitution which guarantees fundamental rights to 'life and personal liberty' to all persons in India is enforceable in a court of law. The Supreme Court has interpreted this article to include the rights of 'living environment', which is implicit in the constitution and therefore, any activity causing damage to environment amounts to violation of article.

Article 253 of the constitution provides provision for making any law. It gives power to the Parliament to formulate laws for the whole or any part of the territory of India for implementing any treaty, agreement or convention with any other country or countries or any decision made at any international conferences, association or other body.

Article 252 (1) of the constitution provides provision for making laws for two or more States on any matters with respect to Parliament has no power to make laws for the States.

The Apex (High Court and Supreme Court) have also accepted Writ petitions relating to environment protection under *Article 32* (Right to Constitutional Remedies). The writ petition is admitted if filed by an alleged sufferer of an action against the individual or company which is causing environmental damage. Besides, its Public Interest Litigations (PIL) can be filed by a sufferer or interested parties (Public minded individual, Voluntary organisations and Judge). Public Interest Litigation (PIL) has originated from the 42nd constitutional Amendment Act, 1976, which introduced *Article 39 A* in the constitution.

Article 39 A emphasises to provide Equal justice and free legal aid to all citizens of India. Even where specific statutes are absent, *Article 372* of the constitution permits application of Law of Trot. This article gives emphasis on the continuance and enforcement of the existing laws and their adaptation in the territory of India until altered or repealed or amended by competent Legislature or other competent authority (The Constitution of India, 1983).

Thus, these constitutional provisions put responsibility on the citizen, State and country for the betterment and cause of environment. These also act to guide in formulation of laws and acts as and when desired by the competent authority for the conservation, protection and preservation of environment and biodiversity.

Acts and Legislations: The Silent Spectators

India has witnessed the enactment legislations directly related to environment. These are:

Water and Air Acts: The water Act, 1974 was passed by the parliament under Article 252 (1) of the constitution of India. Water being a State subject, the act was initially applicable to those states which had passed the enabling legislative resolution under the above article authorising the parliament to enact on a state subject. Subsequently, all other state legislature has adopted the Act in their states at different points of time. The Act has subsequently been amended in 1978 and 1988. The Air Act, 1981 is a central legislation passed by the parliament under Article 253 of the constitution of India. In pursuance of the decision taken by the United Nations Conference on Human Environment held at Stockholm in June, 1972 where India actively participated.

Forest Conservation Act (1980): This is very powerful central Act and conservation of many forests has been possible only because of this Act. According to this, no State Government or any other authority can de-reserve a reserve forest, use any forest land for non-forest purposes or lease a forest of trees naturally grown there for the purpose of using it for reforestation without the approval of the Central Government. Contravention of this by any department of the government or authority will make the head of the government authority liable for punishment or imprisonment up to 15 days.

The Wildlife (Protection) Act (1972): This Act has been enacted by the parliament with a view to protection of wild animals and birds and for matters connected there with or ancillary or incidental thereto.

Environment Protection Act (1986): In 1986 the Union Government enacted the Environment Protection Act to provide for the protection and improvement and prevention of hazard to human beings, other living creatures, plants and properties.

Besides these, the other important prevailing acts are, the Water Cess Act (1977), the Public Liability Insurance Act (1991), the Conservation of Forest and Natural Ecosystem Act (1994) etc. These Acts look for the ways in which pollution can be avoided in future. Similarly, the basic criminal laws of India, namely, Criminal Procedure Code (1898) as revised in 1973 and Indian Penal Code (1860) and Merchant Shipping Act (1958) provide penalty for creating nuisance to water, marine pollution etc.

Though an array of legislations and acts are already made and enacted in order to preservation and conservation of forests, control of pollution and protection of wildlife till today the mission of conservation and protection is not achieved. These legislations and acts have not brought out any appreciable changes in the conservation strategies. It is seen that , stealing of timbers , felling of trees , encroachment of forest lands, poaching and killing of tigers and elephant and incidence of forest fires are occupying regular features in the news headlines of leading newspapers. Even destruction of forest and wildlife is found as a regular incidence. As per example, the wildlife protection act is found being silent so far killing of inland and migratory birds are taken in to account in Nalabana Bird Sanctuary of Chilika Lake in Orissa. Still, the local people are found killing and exporting birds but no exemplary punishment is imposed on the people those are found involved in it. At the national level, killing of black deer by Salman Khan, a film artist though proved legally, but, the Wildlife Protection Act seems to be silent on this matter. Like wise several cases are found pending but the degradation and destruction of environment, forest and wildlife found increasing day by day. Thus, the legislations and acts are found to be mute or silent spectators over the degradation and destruction of forest, environment and wildlife.

Rural Women and Traditional Environmental Wisdom

Rural women have accumulated lot of empirical knowledge on the basis of their experience with nature and natural resources. This traditional wisdom is based on the real realisation that man and nature from part of an indivisible whole. This eco-centric view of traditional societies is widely reflected in their attitude towards plants, animals, rivers and the earth. This whole body of knowledge centered on the economic value of plant and animal species. Rural women hold traditional knowledge in their day-to-day life. In traditional societies people are depend upon tradition for their social, ecological and cultural dimensions. It is seen that, rural women till to day value a number of plant and animal species from the wild for economic, cultural and spiritual reasons. Hindu culture believes in the partnership and stewardship ethics, which stands for holding of land including trees, rivers and mountains, and animals in trust for God and general benefit of mankind.

Traditional medicinal plants are the priceless gift of nature. It is a well-known fact that Indians since time immemorial have been depending on herbal medicine. Rural women till today depend on several plans for their healthcare system. A simple rural woman seems to have applying her own indigenous knowledge for self-cure and curing of her family members. Some common herbs and shrubs like, *Tulsi,* Turmeric, *Adosa, Sant, Pipali,* Ginger, *Brahmi, Til, Giloe, Salsa* etc. are being used by rural women for medicinal purposes in their day to day life.

It may be due to certain benefits, the rural women seem to worship many plants like *Tulsi, Peapal,* Wood apple, Banyan etc., while other plants such as Mango, *Jamun,* Jack fruit, Guava, Palm, and Banana are highly valued for medicinal and fruit production reasons. It is seen that, a rural woman at first applies her own knowledge based natural herbal medicines for the treatment and cure of a disease. Even, sometimes preventive measures are taken by the use of traditional medicine against the outbreak of several diseases. Thus, several plants are found to be intimately linked with culture, traditions and social life of the rural women. The rural women are seemed to protect and plant these plants for medicinal, cultural and ecological values. The culture based Traditional Ecological Knowledge (TEK) is of great beneficial to the society and to the ecological environment as a whole. This knowledge acts as the inter connection element between ecological, socio-cultural and economic dimension of the society. Even these traditional knowledge of the rural women seems too be appreciated now-a-days. At present, more and more people, Government, policy makers and non-governmental organisations are found giving more emphasis on the protection, preservation and conservation of indigenous medicinal plants. Governments both the State and the Central are found giving more emphasis on conservation of these plants through herbal gardens and ex-situ conservation. Thus, the eco-centric, traditional and culture based ecological knowledge of the rural women must be recognised in order to enrich environment end restoration of balance of natural ecosystem.

Environmental Degradation and Women

Women in general and rural women in particular, are vulnerable to negative consequences of environmental production. The reasons are being their heavy workload, poorer health status and tightened household responsibilities. Gathering, sowing, cooking, washing clothes and many more tasks perform by millions of women every single day of their lives. Moreover, when forests recede, it is the women who walk farther for collecting the firewood, and when the water levels fall they walks miles and miles in search of water. Thus, deterioration in the natural resources has direct impact on women's lives. The important sectors where women are affected due to environmental degradation are agricultural production, income generation, nutrition and health of their own family members.

Rural Women in Agriculture

Women perform a critical role in agricultural and live stock production. Both these activities are dependent on the availability and quality of resources like land, water, forests and pastures. Women produce the majority of world's food and contribute 60-80 per cent towards agricultural labourer in Africa and Asia. Women agriculturist is also the decision maker in agricultural production. For example, in Nepal, women take 40-80 per cent of decisions regarding seed selection, use of improved seeds and fertilisers. In developing countries women often farm marginal lands that are subject to decreasing soil fertility, erosion and salinisation (Dash and Mishra, 2001, p.232).

Secondly, women farmers also have limited access to information and technologies for improving the sustainability of agricultural lands because in most developing countries agricultural extension services often do not reach them. As a result women have fewer available substitutes for enriching the soil. For example, when dung and crop residues are used as fuel because of the scarcity of fuel wood, the soil fertility cannot be increased. This leads to significant reduction in agricultural yield. Some times, women receive second hand information, which may be inaccurate and fail to contribute to improve agricultural produce.

Women in Family-Income Generation

Women rely on collection of plants and animals from forests, fresh water and marine habitats for resource based income generation. They collect and process a wide range of non-timber forest products like harvesting fruits, nuts, medicinal herbs, home based processing of foods, fibres, dyes, liquor and making crafts for sale. These are the major source of income for poor women. In India many of the estimated 4 million head loaders of fuel wood are women. Women are also active participants in small scale fishing enterprise and in collection of aquatic and marine resources. (Mishra and Das, 2001, p.82).

Women in Health and Nutrition

Women perform two central activities to ensure health and nutrition of their families i.e. maintaining a clean and hygienic environment and preparing daily food. In order to ensure, that family health is not compromised by unsanitary conditions, women's household activities include washing, mopping, sweeping and removal of faecal matter and household trash. In urban and urban slum areas, the ability to provide a clean and hygienic household is more difficult for women to obtain safe drinking water, prepare and store food in hygienic manner, and remove organic wastes from the vicinity of the household.

The preparation of cooked meals depends on adequate supplies of fuel. In rural and semi urban areas, women and to some extent men and children, collect fuel wood. The time women spend in collecting fuel wood varies according to the availability fuel wood trees, the amount of assistance provided by other members and the household consumption requirements. In India fuel wood collection has been found to require 3 to 5 hours per day and involves travelling distance of 3 to 10 kilometers.

Women in Environmental Protection

Women also participated in some grass root level environmental movement in all over India. With increased environmental awareness, environmental movement is gaining strength day by day all over the World. Today we have green activist in almost every sphere of life.

The Bishnoi community of Rajasthan has a tradition of protecting trees and loving animals, being desert dwellers, they know the importance of trees in the protection of environments. In 1731, the Bishnoi women led by Anita Bai prevented the Maharaja's men from cutting the trees for the construction of the palace by embracing the trees. In this, Anita Bai, her husband and three daughters along with 363 of her followers died in the hands of Maharaja's axe-men. The sacrifice of Anita Bai and her followers is a leading incident in history for the protection of environment. Bachni Devi of Advani village and Gouri Devi of Rani Village drove away the contractor's axe-men in the hills of Uttar Pradesh in

1972. Chamun Devi of Nahikala village in the Doon Valley launched a similar movement in 1986 against a limestone quarry.

Medha Patekar, a renowned social activist led the environment movement against the proposed high dam over Narmada River. She led the movement, the *Narmada Bachao Andolan*, with the help of the tribals whose lands were to be submerged. Although she did not succeed in stopping the project, but the height of the dam was reduced, thereby, decreasing the submergence area. Maneka Gandhi is also a successful environmental journalist. Her writings and TV programmes are very popular, and through these she propagated the cause of environmental protection, love and kindness towards animals, vegetarianism, etc. (Dash and Mishra, 2001, p. 255).

Environmental Enrichment: The Need of the Hour

Environment provides all the necessities for mankind. It is rich in natural resources which are vital for all living organisms, but on the other hand, continuous efforts of the human beings is going on for betterment of life process. The environment plays a very important role not only for the national development but also for the individual development. The impact of environmental pollution on the mental state of an individual in term is reflected in his/her social behaviour or social behavoural patterns. Thus, environmental pollution indirectly contributes towards the degradation social, moral and cultural value patterns of the society. Environmental pollution creates serious health hazards of the citizens and weakens their psycho-social conditions. This ultimately reflected in the national and international scenario. Thus to attain environmental security and to save the entire gio or globe from untimely decay, we should emphasis environmental enrichment programmes like, aforestation, mass plantation, soil conservation, maintenance of hygiene and sanitation, control and minimise pollution, preservation and conservation of forest and wildlife, strong enforcement of acts and laws, better environment management strategies, environmental education and a high level environmental consciousness among all ages and all section of the society including women.

Measures Adopted by Rural Women for Preservation of Environment

Measures adopted by rural women of Tangi block against some areas of environmental degradation are discussed in Table 1.1. For the study 300 rural women were selected by random sampling to collect the opinions through interview schedule.

Table 1.1: Measures Adapted to Check Soil Erosion

Measures	*Frequency*	*Percentage*
Plantation	128	42.67
Embankment	47	15.66
Cruz	15	5.00
Not taken any effort	110	36.67
Total	300	100

It is observed from Table 1.1 that, 42.67 per cent of the rural women used to plant trees, 15.66 per cent of the respondents used to make embankment and only 5.00 per cent of the respondents used to make cruz to check soil erosion. But, 36.67 per cent of the respondents were found not taken any effort to protect and preserve soil.

Due to their strong association with the land or soil, rual women's knowledge with regard to land or soil is unchallengeable. There are instances where such indigenous knowledge beat to proven expertise of skilled engineer and professionals in the field. It is also noteworthy to mention here that, rural women have good knowledge on the role organic materials and fine sediments on cropland. These two in fact are the key components for soil productivity. Therefore, in order to preserve organic materials and fine sediments in the soil rural women are more inspired to resort to the aforesaid concretes measures in order to check soil erosion.

Uninterrupted production and productivity of land is positively co-related with healthy livelihood. In order to maintain such balance rural women never hesitate to execute any work relating to preservation, for enhancing holding capacity and increase productivity of the soil. In order to maintain such capacity of the soil rural women prefer to take measures of their choices suitable for preservation and maintenance of soil quality.

Table 1.2: Measures Taken Against Air Pollution

Types of measures	*Taken*	*Not taken*
Planting trees	187 (62.33)	113 (37.37)
Creating awareness	78 (26.00)	222 (74.00)
Educating children	135 (45.00)	165 (55.00)
Participation in massive plantation	89 (29.37)	211 (70.33)

(Figures in the parentheses are percentage values).

Table 1.2 depicts the measures such as, planting trees, creating awareness, educating children, and participation in massive plantation activities adopted by rural women to check air pollution. Out of the total sample 187 (62.33%) were found planting trees, whereas, 113(37.67%) had not shown any interest in plantation. Air is an extremely important ingredient of the environment required for the sustenance of life on the planet Earth. But, air pollution has become an important causal factor of environmental degradation. As a controlling measure tree plantation is seems to be beneficial as green plants use to take up atmospheric Carbon dioxide for the synthesis of carbohydrate in a complex process called photosynthesis. In this process Oxygen is produced and liberated to the atmosphere. Thus, forests, with lot of plants with them are responsible for the removal of Carbon dioxide from the atmosphere side by side adding Oxygen to the atmosphere. In this regard plants are extremely important for environment and for animals including man.

Out of the total sample, only 78(26.00%) of the respondents were found creating awareness among others; whereas, large share of 222(74.00%) respondents were found away from such task. Though personal involvement and awareness among family members directly influence environmental activities; but in the sample, most of the respondents were not conscious about it. Hence, public, more specifically rural women must be made aware about the pollution hazards and their preventive measures. Similarly, 135(45.00%) of the respondents were found educating their children about the harmful effect of pollution; whereas, 165(55%) did not bother about it. It is seemed to be important that, effects of pollution and consequences of these effects must be understood by all the members of the family and it should be transmitted to the others even from generation to generation. In order to check air pollution, 89 (29.37%) of respondents were found participating in massive plantation activities;, where as, 211 (70.33%) had not shown any interest about this programme. Massive plantation programmes seem to restore ecological balance, maintain biological diversity, helps in water and soil conservation, preventing floods and safe guard future of the tribals and ruralites. Thus, it is seen from Table 1.2 that a good number of respondents had not taken any measures for prevention of air pollution.

Table 1.3 depicts the interest if the respondents in planting different varieties of plants in their residential areas. The researcher had identified five preferred that the rural women had interested in planting i.e. a) flowering and ornamental plant, b) medicinal plants, c) fruit plants, d) forest plants and e) traditional plants. The respondents were asked

Table 1.3: Interest in Selected Types of Plants

Score / *Criteria*	5	4	3	2	1	*Maximum score*
a	05 (25)	10 (40)	01 (03)	21 (42)	05 (05)	115
b	02 (10)	05 (20)	07 (21)	04 (08)	07 (07)	66
c	03 (15)	05 (20)	18 (54)	- -	10 (10)	99
d	06 (30)	03 (12)	05 (15)	06 (12)	03 (03)	72
e	02 (10)	11 (44)	13 (39)	06 (12)	30 (30)	135
Total						487

(Figures in the parentheses are scores).

to give their choice in order of preference. After computing for total responses in five point scale or Likert scale, the maximum scores obtained for interest in planting different plant varieties was 847. Out of the total scores interest in traditional plants obtained a maximum score of 135, whereas interest in flowering and ornamental plants obtained 115 scores followed by fruit plants with 99 scores, forest plants with 72 scores and medicinal plants with 66 scores. Thus 1.3 testify the choice and interest of rural women in planting. It is observed that a significant number of respondents were interested in planting traditional plants, flowering and ornamental plants, and fruit plants. The rural women of Tangi block have a keen interest in planting. They use to plant trees like Wood apple, Mango, Jack fruit Coconut, *Jamun* etc. in their own fields near the houses. Even they plant *Peepal* and *Fucus* trees near the village pond. They also use to plant seasonal vegetables and flower plants near the houses. All the plantation work they generally do with out the knowledge of the male members or when male members are. These activities show their love of nature and contribution for the environment. These also indirectly reflect their love for tradition, culture love for religion.

Table 1.4: Service Rendering for Environmental Enrichment

Type of services	*Frequency*	*Percentage*
Massive plantation	94	31.33
Constructing embankment	52	17.33
Cleaning garbage	112	37.34
Other (if any)	42	14.00
Total	300	100

Table 1.4 reveals the ways that rural women personally involved for environmental enrichment programmes. Out of the total sample 94 (31.33%) of respondents were found involved in massive plantation. They were very much found aware of the benefits of the plants, forest and woodlands. They had also shown their interest about preservation of forests for food and shelter of a wide variety of interdependent animal species in order to maintain biodiversity. No doubt massive plantation is of great significance in soil conservation, checking surface run-off and breaking wind. But for medicinal and religious value point they used to plant some trees in their own gardens as well as near their houses. Similarly, 52(17.33%) of respondents were found to take efforts to construct embankment particularly for conservation of soil. Due to such an effort upper part of soil seems to be protected from erosion and soil fertility is also maintained. Majority of the respondents (37.34% i.e. 112) were found cleaning garbage for maintaining a salubrious surrounding. They were found to clean their own surroundings in order to protect them from pungent smell and to enjoy fresh and free air. Only 42(14.00%) of the respondents were found involved in other activities such as leafy vegetables, creepers and maintaining small kitchen gardens etc. that add to environment in a

specific time or season only. The women in Tangi block were seem to be more conscious about their own surroundings as they were found cleaning weeds around their houses, cleaning garbage, dust and stagnant water near their houses with a view to check mosquitoes, flies and other vectors.

Role of Women in Environmental Protection

Women have very vital role to play in the protection and preservation of environment.

1. As women produce and nurture future generations, they can make their children aware of the preservation of environment.
2. Women as the chief architect of family can make other members of their family learn to make their surrounding clean.
3. A woman can only encourage the rural women. For women, it would be easier to motivate other women.
4. A women by her emotions and feelings can only find out the hidden problems faced by the dwellers in a locality.
5. Women can only discuss about certain sensitive issues like, health care measures and family planning to any locality. Consequently, they can discuss about precaution any and preventive measures.
6. Women located in the rural can conduct seminars and workshop on their problems.
7. Women can organise different campaigns on health care and environmental protection.

9. Women can mobilise funds through voluntary donations for the above activities.

10 Women can attract the attention of media government and International agencies regarding certain developmental activities to be done for sustaining the development of environment (tree plantation,

campaigning for cleanliness of surround, proper waste disposal, use of musk etc.).

REFERENCES

Bandapadhaya, P.K. (1993); Overdrawing from Natures Bank, *Yojana*, Ministry of Information and Broadcasting, New Delhi, 37 (10) pp. 14-15.

Dash, M.C. and Mishra, P.C. (2001); Man and Environment. MacMillan India Ltd., Kolkata.

Gandhi, Indira (1972); First UN Conference on Human Environment, Stockhom.

Ignacimuthu, S.J. (1998); Environmental Awareness and Protection. Phoenix Publishing House, New Delhi.

Mani, J.R. Nair, D. and Ani, J.R. (1997); Medicinal Plants: Our Rich Heritage, *Kurukshetra,* Ministry if Rural Development, New Delhi, Vol. XLVI, No. 3 pp. 29-32.

Mishra, P.C. and Das, R.C. (2001); Environment and Society. MacMillan India Ltd., Kolkata.

Mishra, R.K. (2005); Mother Earth, Daksheshwar Memorial Foundation, Athmallik, Orissa.

Mohanty, P.C. (2003); *Orissa Review,* November, Information and Public Relation Department, Government of Orissa.

Ramakrishnan, P.S. (2001); Ecology and Sustainable Development.

The Constitution of India (1983); Government of India, Ministry of Law, Justice and Company Affairs, New Delhi.

The Samaj (1995); 2nd June, Cuttack.

The Sambad (2005); 7th October, Bhubaneswar, p. 10.

2

Environmental Hazards by Cobalt on Growth and Development of *Lemna minor.* L.

Dr. Minati Mahapatro♣

Dr. L.R. Patro♠, *Dr. B. Padhy*♥

Abstract

By taking different concentrations of Cobalt Nitrate $Co(NO_3)_2$ from 0.50 mg/l to 5.00 mg/l, experiments were conducted to find out the growth and development of Lemna Minor. L. All the test concentrations of $Co(NO_3)_2$ on Lemna Minor L. considerably inhibited the number of fronds, area of the fronds, fresh weight, and dry weight up to 2.50 mg/l. concentration and upto 6th (six) day of treatment and gradually it is decreased with increase in concentration and days of treatment.

Key Word

Phytotoxicity, Auaticangiosperm, Ion, Cobult, Lemna, Pollution.

♣ Laboratory of Physiology and Biochemistry, Department of Botany, Science College, Hinjilicut.

♠ Environmental Toxicology Department, Departmental of Zoology, Kshetrabasi, DAV Science College, Nirakarpur.

♥ Division of Physiology and Biochemistry, Department of Botany, Berhampur University.

Introduction

Sources of entry of metal ions like Cu, Zn, Co, Cd, Ni are through the effluents of textile industries—Srivasthava and Prakash (1991). Vallee and Ulmer (1972) reported that both essential and indifferent metals can be Phytotoxic, when they are assimilated in excess by the plant. In India approximately 27 tonnes of cobalt are being recovered from a lead-zinc ore mine and chromite mines. An important sources of high party cobalt for use in Electrone Industry. (Onsidine, 1974; Morrel, 1979, Donaldson, 1986, Donaldson et al., 1986 a, Budavari, 1989). The waste products of the manufacturing units contain higher level of cobalt which directly or indirectly enter into the environment. The occurrence of cobalt has been reported in air (Elinder and Friberg, 1986; Schrauzer, 1989), tobacco smoke (Iskander et al, 1986), water and sediments (Merian, 1985; Schrauzer, 1989), foods and beverage (Friedrich, 1984; Donaldson et al., 1986 b, Yamagata et al., 1963; Torwitx and Vander Linden, 1974), human tissues and body fluids (Alexanderson 1988; Reynold, 1989, Jyengar & Woittiez, 1988). Cobalt produces toxicity to plants, when the amount exceeds to certain low concentration level. Wallace and Bery (1989) studied the effort of cobalt in Letture seedling grown in culture solution where as Wellace and Abou-Zam Zam (1989) studied the interaction effect of cobalt and Zinc in bush-bean. Dirilzen and Inel (1994) reported the role of cobalt on growth and metal accumulation in Lemna species. *Lemna minor* L., the duck weed of family *Lemnaceae* is an aquatic free floating macrophytes commonly grow in polluted water of ponds, pools and shallow places. Its laxuarent growth depends upon the concentration of polluting chemicals for which attempts are being taken at various levels to assess the amount of polluting substances. Due to its special features, it was used as Bio-indicator to test aquatic pollution. Bishop and Perry, 1991; Kwan and Smith, 1996; Taladsen and Narberking, 1990 and Wang, 1986 and sought for waste water treatment (Zirschky and Reed, 1988).

Materials and Methods

Lemna minor plants were collected from the ponds nearby the Science College, Hinjilicut washed with water thoroughly then disinfected by 70 per cent alcohol followed 1 per cent sodium hypochloride (Naocl) and rinched with distilled water following the method described by (Bowker (1980), then the fronds were grown in culture media in a specially designed culture chamber mentainent at 30 ± 1° C. and provided with 2.8 ± 2 K lux light intensity from the water level from two florescent tube lights in laboratory, photoperiod of 16 Hr. was administrated followed by 8 hour dark.

Gorham's modified culture medium was prepared separately without $Co(NO_3)_2 \cdot 6 H_2O$ which served as solvent for preparation of different concentration of (0.50, 1.00, 1.50, 2.00, 2.50, 3.00, 3.50, 4.00, 4.50 and 5.00 mg/l) of cobalt solutions to get 5.00 mg/l. concentration of cobalt, 0.310 gm. of $Co(NO_3)_2$ was added per litre of cobalt free culture media. Other concentration of cobalt were prepared by diluting the stock solution of 5.0 mg/l of cobalt. Visually selected 20 (twenty) healthy, uniform sised and green coloured fronds were allowed to grow in culture media with different concentration of cobalt in sterilised Petri dishes containing 100 ml. of culture medium/test chemicals and kept in a plant growth chamber provided with light of 2.8 ± 2 K lus intensity from cool white fluorescent tube lights (10 hr. light and 8 hr. dark) at water level and maintained at 30 ± 1°C. The petri dishes were divided into 6 sets, each with 25 petridishes for individual concentration of text chemicals separately and control for growth observation.

The petridishes of individual concentration of test chemicals are used for growth observation of *Lemna minor* L.

At an interval of two days from the date of treatment initiation till 10th day changes in growth characters were recorded.

Results and Discussion

Maximum growth of frond of *Lemna minor* L. were noticed on 6th day of treatment with 2.50 mg/l concentration whereas minimum growth of fronds noticed under same treatment was 27.03 on 2 days after treatment minimum average number of 7.05 ± 1.03 green fronds were noticed in petriplates having 5.00 mg/l concentration of $Co(NO_3)_2$ on 10th day. A positive correlation was noticed between increasing in concentration of test solution and development of number of green fronds and a negative correlation between the same parameters in higher concentration than 2.5 mg/l (Table 2.1 Fig. 2.1).

The maximum area of the fronds also recorded on 6th day of treatment in all treated sets Fig. 2.2. Higher the concentration of test chemical more than 3.50 mg/l, cause significant inhibition on increase of frond area. Maximum average area is 160.03 ± 1.38 mm^2 was recorded. Whereas during the same period of treatment it was 35.03 ± 1.06 mm^2 in fronds treated with 5.00 mm^2 (Table 2.2). On 10th day of treatment average area of green fronds of an 18.01 ± 1.22 in 5.00 mg/l concentration of $Co(NO_3)_2$.

The fresh weight of fronds to different concentration with increase of concentration of $Co(NO_3)_2$ up to 2.50 mg/l and than exhibit a negative correlation. The fresh weight of fronds were found to be higher in both control and treated sets on 6th day of treatment while it was least in fronds treated with 5.00 mg/l concentration on 10th day (Fig. 2.3), the concentrations ranging from 0.50 to 5.00 mg/l exhibited intermediate value between 11.02 ± 1.36 to 0.71 ± 0.08 mg on 10th days of the treatment (Table 2.3 and Fig. 2.3).

Dry weight decreased rapidly (Fig. 2.4), maximum dry weight of 2.63 ± 0.05 mg was recorded in 6th days after treatment with 2.50 mg/l while it was 0.30 and 1.29 mg in fronds treated with 5.00 mg/l of $Co(NO_3)_2$ and control sets respectively. In 10th day the dry weight to be the least (Table 2.4).

Table 2.1

Treatments	*Number of green fronds per petridish*				
Concentrations of $Co(NO_3)_2$ in mg/l	*Days after treatment*				
	2	*4*	*6*	*8*	*10*
Control	21.02±1.82	23.00±1.55	26.51±2.52	23.52±2.24	22.0±2.23
0.50	22.01±2.02	24.31±2.34	28.33±3.03	25.05±1.54	23.21±1.31
1.00	23.03±1.33	26.05±1.45	30.02±2.24	28.01±3.22	24.03±2.23
1.50	25.05±1.43	27.02±2.14	32.52±1.36	30.51±1.62	25.42±2.46
2.00	26.06±1.12	28.17±2.22	34.51±1.34	31.01±3.12	26.71±1.45
2.50	27.03±1.25	29.21±1.56	35.04±2.34	32.02±3.31	28.22±2.24
3.00	25.51±2.54	26.52±2.44	29.72±1.91	28.13±1.86	26.62±2.12
3.50	22.82±2.81	25.13±2.31	27.04±2.12	24.52±1.24	23.02±1.31
4.00	20.41±2.34	22.71±2.62	21.05±1.25	18.21±2.82	17.03±3.15
4.50	18.31±1.83	17.22±2.62	14.31±2.33	12.52±1.54	11.02±2.03
5.00	16.22±1.23	12.81±2.27	10.03±1.24	8.51+1.52	7.05±1.03

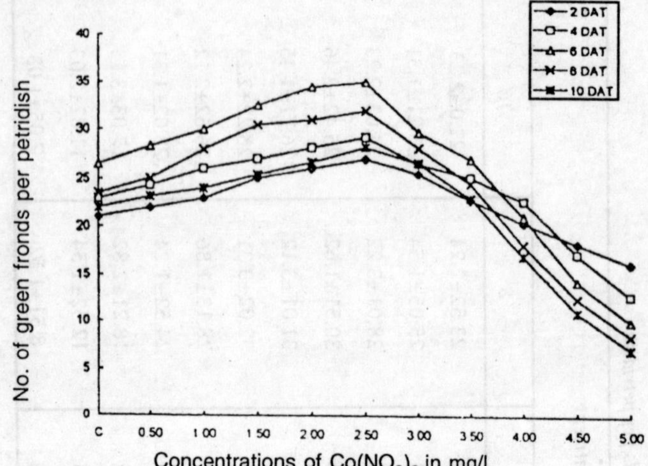

Fig. 1.1

Effect of Different Concentrations of $Co(NO_3)_2$ on Growth of Lemna Minor Fronds.

***C*-Control, *DAT*-Days After Treatment**

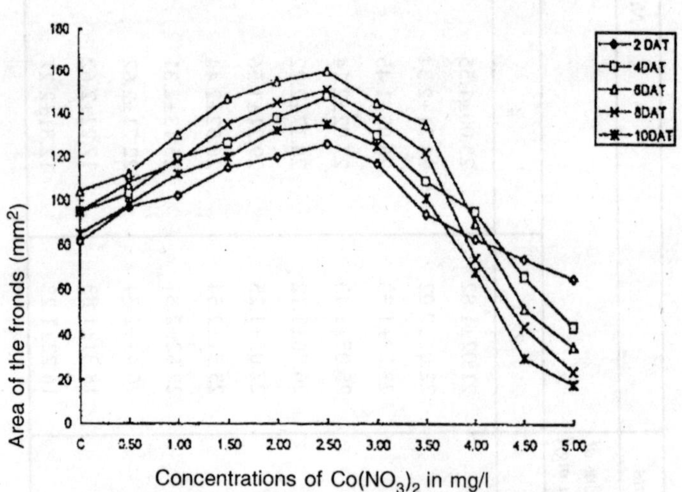

Fig. 2.2

Effect of Different Concentrations of $Co(NO_3)_2$ on Area of the Green Fronds of Lemna Minor Fronds.

***C*-Control, *DAT*-Days After Treatment**

Table 2.2

Treatments	*Area of the green fronds per petridish in MM^2*				
Concentrations of $Co(NO_3)_2$ in mg/l	*Days after treatment*				
	2	*4*	*6*	*8*	*10*
Control	81.54±1.52	94.82±2.02	104.03±1.04	95.02±1.41	85.01±2.11
0.50	97.08±1.34	103.09±1.14	112.01±1.22	108.03±2.24	98.02±1.12
1.00	102.08±3.22	119.03±2.24	130.02±1.42	118.08±2.33	112.03±1.72
1.50	115.09±2.24	126.08±1.33	147.03±2.04	135.02±2.05	120.08±1.63
2.00	120.02±3.31	138.03±2.09	155.08±1.18	145.08±1.13	132.09±2.24
2.50	126.02±2.26	148.08±2.29	160.03±1.38	151.01±1.46	135.04±1.49
3.00	117.06±1.18	130.05±1.09	145.03±2.01	138.04±2.52	125.05±1.13
3.50	94.02±1.12	109.08±3.02	135.06±2.62	122.05±3.03	101.05±2.54
4.00	83.08±1.06	95.08±1.41	90.09±2.23	74.04±2.14	68.03±1.52
4.50	74.02±1.34	66.06=1.37	52.04±1.49	44.03±2.43	30.04±2.62
5.00	65.02±1.34	44.05-1.51	35.03±1.34	24.02±2.15	18.01±1.22

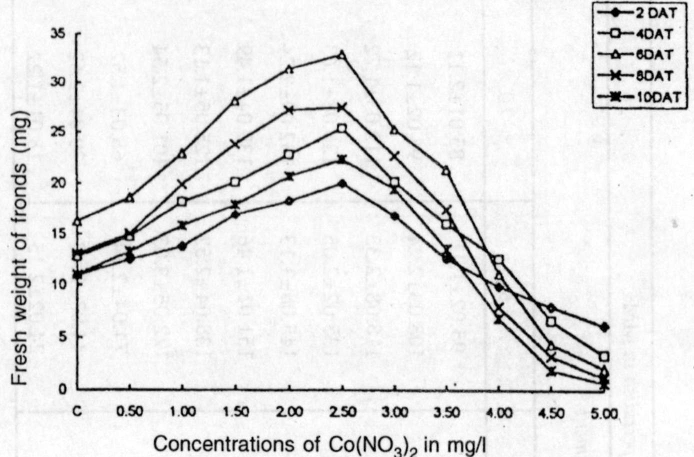

Fig. 2.3
Effect of Different Concentrations of $Co(NO_3)_2$ on Fresh Weight of Lemna Minor Fronds.
***C*-Control, *DAT*-Days After Treatment**

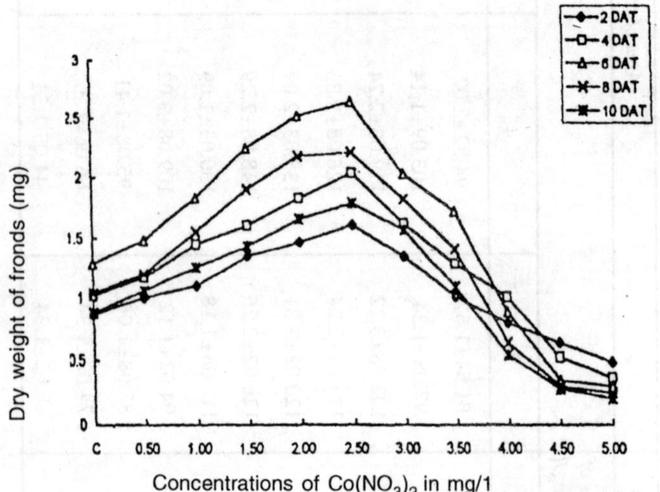

Fig. 2.4
Effect of Different Concentrations of $Co(NO_3)_2$ on Dry Weight of Lemna Minor Fronds.
***C*-Control, *DAT*-Days After Treatment**

Table 2.3

Treatments	*Fresh weight of the fronds in mg*				
Concentrations of $Co(NO_3)_2$ in mg/l	*Days after treatment*				
	2	4	6	8	10
Control	10.92±2.21	12.81±3.42	16.23±1.35	13.11±2.56	11.02±1.36
0.50	12.53±3.22	14.72±2.67	18.61±3.82	15.03±1.72	13.31±2.66
1.00	13.81±2.75	18.20±3.54	22.91±1.82	19.93±2.34	15.82±1.65
1.50	16.92±1.91	20.12±1.26	28.13±3.89	23.81±2.18	17.93±1.47
2.00	18.31±3.43	22.83±1.74	31.41±1.99	27.22±2.31	20.71±1.25
2.50	20.05±1.02	25.41±1.32	32.93±2.87	27.52±2.45	22.34±1.33
3.00	16.81±1.21	20.23±1.22	25.34±1.23	22.73±1.72	19.51±2.11
3.50	12.64±3.24	16.05±1.06	21.42±2.26	17.51±1.35	13.62±2.07
4.00	9.91±0.89	12.63±13.26	11.12±1.22	7.29±0.72	6.81±0.91
4.50	7.93±0.82	6.62±0.51	4.35±0.24	3.21±0.54	1.92±0.86
5.00	6.16±0.15	3.32±0.21	2.06±0.07	1.22±0.09	0.71±0.08

Table2. 4

Treatments	*Dry weight of the fronds in mg*				
Concentrations of $Co(NO_3)_2$ in mg/l	*Days after treatment*				
	2	4	6	8	10
Control	0.87±099	1.02±0.07	1.29±0.08	1.04±0.09	0.88±0.05
0.50	1.00±0.06	1.17±0.08	1.48±0.05	1.20±0.07	1.06±0.02
1.00	1.10±0.04	1.45±0.01	1.83±0.04	1.55±0.04	1.26±0.05
1.50	1.35±0.08	1.60±0.04	2.24±0.09	1.90±0.03	1.43±0.21
2.00	1.46±0.02	1.82±0.08	2.51±0.04	2.17±0.01	1.65±0.04
2.50	1.60±0.01	2.03±0.05	2.63±0.08	2.20±0.02	1.78±0.02
3.00	1.34±0.07	1.61±0.02	2.02±0.04	1.81±0.06	1.56±0.01
3.50	1.00±0.07	1.28±0.01	1.71±0.06	1.40±0.07	1.08±0.03
4.00	0.79±0.08	1.00±0.02	0.88±0.06	0.63±0.02	0.54±0.05
4.50	0.63±0.07	0.52±0.06	0.34±0.09	0.30±0.09	0.28±0.07
5.00	0.48±0.02	0.36±0.90	0.30±0.05	0.25±0.06	0.21±0.02

Conclusion

From the above it concluded that the cobalt and its compounds significantly effect on plants, animals and even human beings. From the growth performance of *Lemna minor* L., Lemna is very sensitive plant to any sort of pollution in an aquatic ecosystem.

Acknowledgement

Authors wish to thanks the Head, PG Department of Botany, Berhampur University providing necessary laboratory facility, Head of the Department, Botany and Major R.R. Padhy, Principal Science College, Hinjilicut for giving the encouragement.

REFERENCES

Vallee, B.L. and Ulmer, D.D. (1972). Biochemical Effects of Mercury, Cadmium and Lead. Annual Rev. of Biochem., 41:91-128.

Morral, F.R. (1979). Cobalt Compounds in: Kirk—Othmer Encyclopaedia of Chemical Technology, (eds, Mark, H.F., Othner, D.F.; Overberqer, C.G; Seaberq, G.T. and Grayson, M.) 3rd ed., Vol.6, John Willey & Sons, Publishers, New York, pp. 495-510.

Donaldson, J.D. (1986). Cobalt and Cobalt Compounds. In: Glman's Encyclopaedia of Industrial Chemistry (eds. Gerhartz, W., Yamamoto, Y.S.; Campbell. F.T., Pfee Fferkorn, R and Rousa Vile, J.F.) 5th ed. Veinheim, VCH—Verdag, pp. 281-313.

Donoldson, J.D.; Clark, S.J and Grines, S.M. (1986 b). Cobalt in Medicinal, Agriculture and the Environment. Reports Slough Cobalt Development Institute.

Budavari, S. (1989). In: the Merk Index (ed. S. Budhavari) 11th ed., Rahway, N.J. Merck & Co., pp. 379-382.

Elinder, C.G. and Friberg, L. (1986), Cobalt. In: Handbook on the Toxicology of Metals. (eds. Friberg, L., Nordberg G.F. and Vouk, V.B.) 2nd ed. Emsterdam, Elsevier Publishers, pp. 211-232.

Sehrauzor, G.N. (1989). Cobalt. In: Metals and their Compounds in the Environment, Occurance, Analysis and Biological Relevances (ed. Merian, E. Weinhein) VCH—Verlag. Publishers Germany, pp. 2-8-1-2-8-11.

Iskander, F.Y.; Bauer, T.L. and Klein, D.E. (1986) Determination of 28 Elements in American Cigarette Tobacco by Nuetron—Activation Analysis. Analyst; 111: 107-109.

Merian, E. (1985). Introduction on Environmental Chemistry and Global Cycles of Chromium, Nickel, Cobalt, Beryllium, Arsenic, Cadmium and Selenium and their Derivatives. In: Carcinogenic and Mutagenic Metal Compounds. (eds. Merian, E.; Friem, R.W.; Hardi, W. and Schlatter, C.), Gordon and Breach Publishers, New York, pp. 3-32.

Friedrich, W. (1984). Vitamins. In: Kirk-Othners Encyclopaedia Chemical Technology, (eds. Mark, H.F., Othmar, D.F.; Overberger, C.G.; Seaverg, G.T. and Grayson, M.) 3rd cd. Vol. 24, John Wiley & Sons Publishers, New York, pp. 158-185.

3

Assay of Total Protein Content in Kidney Tissue of Albino Rat, *Rattus norvegicus*, with Effect of Neem Seed Kernel Extract

*Surendra Singh**
*Rajni Chauhan**

'Neem' *Azadirachta indica* has been called as "Heal all", "Divine tree", "Village pharmacy" and even "Nature's drugstore". Polyhedral preparations containing one, two or all five parts of the plants i.e. leaves, bark, flower, fruit and root, called (panchang) in Ayurveda, were used. It was well known to Indian farmers that locusts during invasion avoid the neem tree and that it has an antifeedant property. Pradhan and Jotwani (1962, 1968) brought this fact to the notice of the scientific world. Different parts of this plant have been reported to have antiseptic, wound healing, skin disease curing and antiulcer activity (Kirtikar and Basu, 1933, Chopra *et al.*, 1956).

Adult albino rats, *Rattus norvegicus, of* both sexes, weighing 120-180 g were inbreed in the laboratory of

* Department of Zoology, School of Life Sciences, Dr. B.R Ambedkar University, Agra-282002 (U.P.).

Zoology Department, School of Life Sciences, Khandari Campus, Dr. B.R. Ambedkar University, Agra and maintained at 25±0.5°C in well ventilated polypropylene cages of 45cm × 27cm × 15cm under a relative humidity of 60±5 per cent and a photoperiod of 12 hrs per day. Neem seed kernel extract was prepared in the laboratory of different concentrations.

The rats were divided into nine groups i.e. A, B, C, D, E, F, G, H and I. The group name A, D and G is the control set of rats for 30 days, 60 days and 90 days respectively. Group B, E and H received 0.8 per cent concentration for 30 days, 60 days and 90 days respectively. The last group C, F and I received 1.6 per cent concentration of the neem seed kernel extract for 30 days, 60 days and 90 days respectively. The calculation of ED_{50} is done by Thompson and Weil method (1952). The albino rats were dissected carefully and kidney was taken out for biochemical estimation of total protein content by Lowry *et al.* (1951) method.

The values obtained were manipulated by F-test (ANOVA) and then after Student's 't' test.

In our study, the ED_{50} value for neem seed kernel extract is 1.1 mg/kg/body weight, while a different ED_{50} value for albino rat i.e. 100 μg/kg/body weight was reported by Hullihan *et al.* (1990).

Kidney is a major site for the metabolism of many small molecular weight proteins and polypeptides, including many hormones such as insulin, PTH and calcitonin. The increase in the mean value of kidney total protein is highly significant ($P<0.01$) after 30 days and very highly significant ($P<0.001$) after 60 and 90 days.

The increase in kidney total protein from amino acid causing a declination of amino acid pool. This increase might be for energy source of spare during chronic condition of stress or drugs. Substances present in neem seeds such as proto-meliacins, meliacins, pentanortriterpenoids, hexanotriterpenoiud and a variety of non-suspenoidal

Table 3.1: Kidney Total Protein of Albino Rats After Treatment With Neem Seed Kernel Extract

No. of Albino rats	*Treatment Days*	*Control set*	*Experimental sets Dose concentrations*		*F-Value*
			0.8%	*1.6%*	
		Range Mean±S.Em	*Range Mean±S.Em*	*Range Mean±S.Em*	
6	30	130.2-132.6 131.35±0.43	133.8-134.5 134.08±0.11	133.5-135.1 134.7±0.24	F=36.82
6	60	131.3-132.4 131.97±0.20	134.3-135.0 134.75±0.11	133.2-136.2 135.42±0.47	F=133.2
6	90	130.3-133.4 131.43±0.48	136.0-136.8 136.35±0.14	137.0-139.2 138.3±0.34	F=103.04

components and carbohydrates (Wealth of India, 1985) effect the change in kidney total protein. Uterine protein is increased in findings of Pathak *et al.* (1992) after treatment *of Ferula jaeschkeana* in rats, while Meenakshi *et al.* (1995) observed a decrease. Also, Alam *et al.* (1992) reported decrease in protein content after treatment of *Vicoa indica.*

Further, the increase in total protein is higher for higher concentration of the extract and for a longer period i.e. 90 days. This shows the dose dependent and time dependent effect of the extract (Prashad *et al.*, 1994).

REFERENCES

Alam, M., Susan, T., Joy, S. and Kundu, A.B. 1992. Anti-Inflammatory and Antipyretic Activity of Vicolides of *Vicoa Indica* DC. Indian J. Exp. Biol.; 30(1): 38-41.

Chopra R.N., Chopra, I.C., Handa K.L. and Kapur, L.D. 1958. Indigenous Drugs of India, 1.

Kirtikar, K.R. and Basu, B.D. Indian Medicinal Plants. 2nd eds. Allahabad, Lalit Mohan Basu 1933; 452.

Lowry, O.H.N., Rosebrough, N.J., Farr, A.L. and Randall R.J. 1951. Protein Measurement with the Folin Phenol Reagent. J. Biol Chem., 193: 265-275.

Meenakshi, C., Latha Kumari, K. and Shymala Devi, C.S. 1995. Biochemical Studies on the Effect of S-1, 3-butanediol of Diabetes Induced Drats. Ind. J. Physiol. Pharmacol., 39(2): 145-148.

Prashad, O., Singh, P., Gardener, P., Fletcher, C.,' Richards, E. and Choo-Kang, E. 1994. Effect of Aqueous Neem *(Azadirachta Indica)* Extract on Testosterone and Other Blood Constituents in Male Rats.

Weil, C.S. 1952. Table for Convenient Calculation of Median Effective Dose (LD_{50} or ED_{50}) and Instruction in their Use. Biometrics, 8: 249-263.

Ayruvedic and Economic Importance of Forest
(Case Study Kondhamal District of Orissa)

Dr. Ram Prasad Panda♣
Ashok Kumar Shadangi♠

Ethonobotanical researches have brought many new faces to the limelight which were unknown till the recent past and opened new avenues of researches in the field of medicines, drug explorations and on feasibility of setting up cottage industries based on forest products and various other socio-economic aspects.

Introduction

India, by virtue of its varied climatic conditions is unique and bestowed with enormous biological diversity. These biological resources are the main sources of food, fuel and fabrics and medicines. Andhra Pradesh is the fifth largest state in India and consist of 8 per cent of the total land area of the country. It has about 65,000 km^2 of forest and ranks fourth in the country's state forest area. About 30 per cent of the states forest area with good biological resources is

♣ H.O.D. in Zoology, Anchalik Science College, Kshatriyabarapur, Ganjam.
♠ Lecturer in Botany, K.S.U.B. College, Bhanjanagar, Ganjam.

located in the Northern district (Lakshimi Narayana & Venkaiah, 1998).

Ayurvedic system of medicine which claims its origin from the Vedas was developed into an elaborate science as is evident from the *Charaka Samhita* and *Sushruta Samhita* written not later than 1000 B.C. These two works give a detailed description of medicinal plants with their uses. Plants have been given specific names and are often grouped on the basis of their curative actions. (Anonymous, 1976, Arulmozhiyan, Salvary Vijaylakshimi, Dhanjayana & Jehangir 1992.)

India has about 100 types of oilseed-bearing perennial trees, mostly grown in forests. Amongst these a large number of trees are identified for commercial exploitation. A few multipurpose trees deserve immediate attention in the aforestation programme in the Kondhamal with varied climates and soils, not only for the conservation of eco-restoration and sustainable development in the region but also for their commercial exploitation.

Study Area: Kondhamal district of Orissa (Circa 19°34′ and 20°54′ N and 83°30′-84°E) is bound in the north by Boud district, west by Bolangir and Kalahandi South by Koraput and East by Ganjam district. This area is dominated by mixed forest. Tribal constitute about 69.59 per cent of the total population.

Results and Discussion

During the field studies in hilly areas of the district several areas of such deserted land which were under′ podu′ cultivation, have been recorded along the various forest ranges.

Based on floral forms, pollination mechanisms and richness of floral food, the bees employ different foraging behaviours for efficient exploitation of floral recurred. In doing so the flowers get pollinated, which is essential for sexually reproducing plant species. Some outcrossing plant species have obligately adapted to pollination by carpenter bees by protecting the nectar chamber against perching from outside, providing a spacious floral tube, and requiring mechanical force to reach nectar. Some others have adapted to pollination by carpenter bees along with other insects. Still some others have developed mutual relationships with carpenter bees. Such a co-ordinated interaction in plant and bee partners benefits each other, the former for pollination and the latter for food therby allowing both to flourish in the habitats where they occur.

Kondhamal forest is rich sources of medicinal plants. Urani and Ayurvedic medicines are prepared from various parts of many wild plants, a few allopathic medicines are also obtained from plant extract. Some important drugs obtained from roots, leaves, woodbark, flowers, fruits and seeds.

The vegetation of the area is now highly disturbed and many of the herbal species are depleting at a rapid rate, Review of literature reveals that the work of Fyson (1915) and Gamble (1915-36) are the only published account of the plants of this area. Since then no attempt has been made to present a picture regarding the herbal diversity and their utilisation. Hence, the present work is aimed at to bring into light the valuable herbal wealth of Ootacamund (Parvathi & Banumathy 1998).

Kondhamal, a vast land of forests, 5th of its kind in the country is inhabitated by varieties of trees, herbs, animals and birds.

It yields a number of minor and major forest products. The major products are timber and firewood. These are used for industrial and fuel purposes. A part from providing raw materials for pulp, panel products, match wood and other wood based industry, forest are also a source of a number of useful minor products like bamboo, canes, kendu leaves, grasses, essential oils, medicinal plants, lac, resins, fatty oils and fats, gum, tanning materials dyes, animal products etc. Quite a number of these are valuable foreign exchange earners.

The tribals living in the forest have been using the plants for their food, medicine, clothes and their necessities of life since time immemorial. We could realise the importance of their extensive knowledge about the easefulness of plants only during the last few decades. Today, the people of different aptitudes, (Botanists, drug explorers, industrialists, naturalists etc.) are all running to them to know about the utility of plants which have formed a part of scientific study today known as ethnobotany which deals with the study of plants in relation to tribal races. Ethonobotanical researchers have brought many new faces to the limelight which were unknown till the recent past and opened new avenues of researches in the field of medicines, drug explorations and on feasibility of setting up cottage industries based on forest products and various other socio-economic aspects.

Kondhamal forests are rich in drugs of immense value. Therefore, the area should be thoroughly explored and the plants of medicinal utility should be sorted out as per the informations gathered from tribal people. It is unfortunate that destruction of forest going is ruthlessly and many of the medicinal plants are on the verge of extinction now. In order to save these plants from further depletion, conservation has become a necessity today. Since these plants are medicinally important for various diseases of heart, gynic trouble, gout, rheumatism, dysentery, diarrhoea, leprosy, bone fracture, skin diseases etc., their preservation and propoagation should be done and for that a medicinal garden should be maintained in Kondhamal.

All these plants may need to be phytochemically investigated and various drugs extracted from them proved to be beneficial to the community.

Considering the high prices of drugs and medicines in the country and our greater independence on other countries for more to it, Scientists and medicines, besides manufacturing synthetic drugs, should also look to the natural drugs obtained from plants. Daringbadi has vast resources of potential drugs of great theraputic value. Search

Table 4.1: Some Common Medicinal Plants of Daringbadi Forest and their Uses

Sl.No.	*Botanical Name*	*Oriya Name*	*Use*
1.	Abrus precatorices	Koincho	Root is used in cold, cough, asthma, bronchitis, as antiseptic and as blood purifier
2.	Achyranthus aspera	Apamarango	Root is used as an antidotes to snake-bite and scorpion sting, in cold, cough, asthma and bronchitis and plant for headache
3.	Adina cordifolia	Kurum	The shoot apex is used in ophthalmic infection and as veterinary medicine
4.	Aeql emarmelos	Bela	Fruit and root used for gastric disorders and leaves for jaundice
5.	Aerve lanata	Jata jatia	Plant paste is used on cuts, wounds and bums
6.	Agave americana	Baramasi	Leaf paste is applied on cuts, wounds and burns.
7.	Ageratum comyzodes	Poko-sunga	Leaf is applied on cuts, wounds, burns and various skin diseases
8.	Alangium salvifalium	Ankulo	Stem bark is used for abortion and menstrual disorders
9.	Alstonia scholaris	Chhatiana	Stem bark is used in viral fever and malaria and its latex (milky juice) in opthalmic infections
10.	Alysicarpus vaginalis	Ratani	Root shows antifortility effects
11.	Amarangthus spinocus	Kanta-Neutia	Root is used as an antidote to snake bites and scorpion stings

(Contd....)

Table 4.1: (Contd....)

Sl.No.	*Botanical Name*	*Oriya Name*	*Use*
12.	Andrographis paniculata	Bhuin-Nimbo	Leaf is used in Malaria fever, also as antiseptic and blood purifiers and the stem in gastric disorders
13.	Annona sgumosa	Ata	Seeds used for abortion and in menstrual disorders and thalamus for antifertility.
14.	Argemene maxicana	Agara	Leaf is used in cuts, wounds, bums, boils, blisters and ulcers and latex (Yellow milky juice) in ophthalmic infections.
15.	Aristolochia indica	Rudra-Iswar Mula	Jato or root is used as antidote to snake bite and scorpion stings stimulant and also cures syphites and fever etc.
16.	Asparagus racemosus	Satabari	Plant is used in headaches.
17.	Azadirachla indica	Neem	The plant has antihelmintic and insesticidal properties. Young leaves and flowers are fried in mustard oil and eaten for skin diseases.
18.	Bauhinia acuminata	Kanchana	Stem bark is used as antidote to snake bites and scorpion stiring.
19.	Borreria articularis	Jibho-kata	Root is used as antidote to snake bites and scorpion stings and plants for pains, swellings and body aches.
20.	Bridelia airyshawii	Kosha Koli Koshi	Stem bark decoction with country Gochha liquor is given after menstruation to prevent pregnancy

Table 4.1: (Contd....)

Sl.No.	Botanical Name	Oriya Name	Use
21.	Calotro pisgigantea	Arakha	Latex used for hydrophobia.
22.	Canscora diffussa	Sankha-puspi	Plants used in headaches, malaria fever.
23.	Capparis zeylanica	Asdhua	Plants shows antifertility effects.
24.	Cassia fustula	Sunari	Seeds shows antifertility and cooling effect, also used in gastric disorders.
25.	Cussia occidentalis	Chakunda	Root is used in gastric disorders and leaves and seeds in skin diseases.
26.	Cissampelos pareira	Okanbindhi	Root is used in gastric disorders, malaria and fever.
27.	Cochlospermum gossypium	Dalina	Stem bark is used in jaundice. Gum used as sedative.
28.	Cocculus hirsutus	Musakani, Dahadahiya	Root is used in rheumatism, gout and veneral diseases.
29.	Colsbrookea oppositifolia	Bhainsa	Leaf is used in cuts, wounds and burns and roots in epilepsy and hysteria.
30.	Combretum roxburghii	Atundi	Leaf shows antifertility effect.
31.	Curcaligo orchioides	Talmuli	Root used as antidote to sanke bite and scorpion sting.
32.	Cuscuta reflexa	Nirmuli	Plant is used in cuts, wounds, burns and various skin diseases.
33.	Lypersus totundus	Motha	Bulb is used in gastric disorders.
34.	Dalbergia sisso	Sisu	Leaf and root used for gonorrhoea.

(Contd....)

Table 4.1: (Contd....)

Sl.No.	*Botanical Name*	*Oriya Name*	*Use*
35.	Desmodium gangeticum	Salaprani	Root is used as an antidote to snake bite and scorpion strings and plants in ophthalmic infections.
36.	D.triflorum	Kurudia	Plant is used in pains, swellings gachha and bodyache.
37.	Eclipta prostrata	Keshadura	Root is used as antidote to snake bite and scorpion strings and leaf is used in fever and malaria.
38.	Fuphobia hirta	Bula siju	Leaf is antidote to snake bites and scorption strings and latex used for toothache.
39.	Flemingia strobilifera	Harabhanga	Root is used in epilepsy, hysteria and gastric disorders.
40.	Gordenia gummifera	Kuruda	Plant is used for bone-fractures.
41.	Gmelina arborea	Gamhari	Leaves used as an antidote to snake bite and scorpion string and root for malaria and fever.
42.	Grewiao btusata	Asana gochha	Plant is useful in boils, blisters and ulcers.
43.	Helictresisora	Atmura mudimudica	Bark is useful in dysentery and diarrhoea root in diabetes and fruit in griping bowels. Fruits decoction in fever.
44.	Hemidesmus indicus	Ananta Mula	Root is useful in impotency and other sexual disorders and various skin diseases.
45.	Hoppea dichotoma		Plant is useful in bone-fracture.

Table 4.1: (Contd....)

Sl.No.	*Botanical Name*	*Oriya Name*	*Use*
46.	Ichnocorpus frutescens	Tapar-hahar, Soyam Nai	Plant stem is useful in toothache.
47.	Indigofera linaei		Root shows antifertility effect and juice is used as alterative and diuretics.
48.	I-tinctoria	Neli	Plant extract is used in epilepsy, nervous disorders, in bronchitis and as on ointments in sores, ulcers and root in hepatitis.
49.	Leobotis nepetacfolia	Kantisia	Flowers and seeds are used in cuts, wounds and burns, leaf decoction as tonic.
50.	Modhuca longifolia	Mahul	Bark power is used in toothache and tonsillitis and its decoction as tonic, flowers fried in ghee is eaten for piles.
51.	Mallotus philippinensis	Kapala gunda	Seeds useful in boils, blisters and ulcers.
52.	Mangifera indica	Amba	Fruit is rich in Vit. A, C and D when unripe useful as opthlmic and in eruptions and seeds in asthama.
53.	Millettia pinnata	Karanj	Seed oil is useful in cutaneous infections herpis and scabies and fresh bark in bleeding piles.
54.	Mimsa rubicaulis	Ralo	Leaf is used in epilepsy and hysteria
55.	Morings oleifera	Sajana	Stem bark is useful in baldness.
56.	Mucuna pruriens	Baidanga	Leaf is used in boils, blisters and ulcers, seeds as nervine tonic, pody (fruits) as anthelmintic and infustion of root mixed with honey is given in cholera.

(Contd....)

Table 4.1: (Contd....)

Sl.No.	*Botanical Name*	*Oriya Name*	*Use*
57.	Nelumbo nucifera	Padma	Flower is useful in diarrhoea, in fever, liver trouble, in cholera, as cardiac tonic, root in piles, dysentery and rhizome paste in ring worm and skin diseases.
58.	Ochna obtusata	Bhuin cham Bhuin Kaniara	Plants used in veneral diseases.
59.	Pedilanthus ththy maloides	Khadi siju	Latex is useful in veneral diseases.
60.	Pergularia daemia	Itudrudi	Leaf juice is useful in catarrhal affections, infantile diarrhoea and asthma, leaf paste as poultice on carbuncles.
61.	Plumbgo rosea	Rakta-chita Lal chandua	Stem latex is applied in head to cure headache
62.	Richinus communis	Jada	Leaf is used as poultice for boils.
63.	Rotala densiflora	Roth	Plant is used for deafness.
64	Ribis cordifolia	Alti	Root decoction is used in urinary trouble. Stem in cobra bite and scorpion sting.
65.	Rauxolfia serpentina	Patal garuda	Pain killer, Leaf juice is blood purifier
66.	Semicorpus anacardium	Bhalia	Nut oil is anti cancerous, externally applied in rheumatism and leprous nodules, dried thalamus'used as tonic after delivery.
67.	Smilex macrophylla	Muturi Gachha	Plant is used in abortion and menstrual disorders.
68.	Sphaeranthus indicus	Mundi	Leaf juice is given in malaria and fever, in liver and gastric disorders and with boiled mild in cough.

Table 4.1: (Contd....)

Sl.No.	*Botanical Name*	*Oriya Name*	*Use*
69.	Syzygium cumini	Jamun	Aquous extract of seeds or powder is very much helpful in decreasing blood sugar
70.	Terminalia arjuna	Arjuna	Bark extract is used in heart diseases, as cardic tonic for sores and as an antidote to poison.
71.	Tinospora cordifolia	Guluchi	Plant is used for bone fractures.
72.	Toona ciliata	Maha-Nimbo	Stem bark is used in gastric disorders.
73.	Tridex procumbens	Bislyakarani	Leaf is used in ophthalmic infections.
74.	Ventex negudo	Begunia Nirgundi	Leaf is used in pains, swellings and body-aches.
75.	Yeromphis uliginosa	Mohana	Root is used in deafness, malaria and fever.
76.	Glorisa superba	Pancha	Plant produces an alkalaid which Angulia cures gout.
77.	Albizzia lebbeck	Seris	Used in night blindness.
78.	Flacourtia indica	Baincha koli	Used for jundice and enlarged spleen
79.	Martynia annua	Baghnakhi	Used for epilepsy.

Table 4.2: Selected Trees and their Use

Sl.No.	Botanical Name	Oriya Name	Uses
1.	Acacia catchu	Khaira	Used in pan, medicinally for sore throat, cough and pest
2.	Acacia leucophlaea	Gohira	Agricultural implementation, oil mills, fibre, gum, bronchitis.
3.	Adina cardifolia	Halanda	Flooring, railway, carriage, plywood, fodder.
4.	Aegle marmelos	Bela	Diarrhoea, dysentery, laxative, diuretic, gum, oil.
5.	Ailanthus excelsa	Maharim	Asthma, bronchitis, dysentery, fodder.
6.	Albizia lebbeck	Kala sirisa	Furniture, flooring, railway carrage, gum, tanning, eye trouble.
7.	Albizia orocero	Dhala sirisa	Chronic diarrhoea, asthma, beriberi, congested liver, tea boxes, writing board etc.
8.	Alstonia scholaris	Chatiana	Choronic diarrhoea, asthma, beriberi congas liver, tea boxes, writing board etc.
9.	Alstonia venenta	Ranajita	Mental disorder, alkaloids
10.	Angeissus latifolia	Dhaura	Gum pharmaceutical prep, poles, agricultural implements and fumiture.
11.	Anthosephalus cademba	Kadamba	Construction work, flower receptacle edible, tonic.
12.	Ardisia solanceae	Batguri	Leaves as salads, fever, diarrhoea rheumatism and dying.

Table 4.2: (Contd....)

Sl.No.	*Botanical Name*	*Oriya Name*	*Uses*
13.	Azadirachta indica	Nimba	Skin trouble, antiseptic, washing soap, oil, timber etc.
14.	Barringtonia acutamgula	Hinjal	Leaves edible seed is emetic and expectorant, bitter tonic, tannin, boat building.
15.	Bassia latifolia	Mahul	Oil, edible, liquers, vinegars, tonic, musical instruments.
16.	Bambura aurndinaca	Baunsa	Paper industry.
17.	Bombax malabaricum	Similitula	Match industry, pencil, plywood insulating material for refrigerator, tonic, diuretic.
18.	Barasus flabellifera	Tala	Leaves as writing material, fan, thatching, mats, buckets.
19.	Bridelia retusa	Kasi	Agricultural implements, tool handle, antiviral, hypoglycemic, hypotensive, fodder etc.
20.	Buchanania lanzan	Charkoli	Glandular swelling of reck, beam, window frame, match boxes, dressing textiles etc.
21.	Butea monosperma	Palasa	Making plotters, ropes, piles, tumours, menstural disorder.
22.	Boswellia serata	Salai	News print, typopaper, oil, turpentine paints, textile industry, diarrhoea, skin trouble etc.
23.	Careva arborea	Kumbhi	Agricultural implements, cabinets, paper making.
24.	Casia fistula	Sonari	Tanning, purgative, laxative, tool handle.

(Contd....)

Table 4.2: (Contd....)

Sl.No.	Botanical Name	Oriya Name	Uses
25.	Casuarina equisetifolio	Jhaun	Fuel, fencing, diarrhoea, dysentery, colic, dying, tanning, resin.
26.	Cleistanthus collinus	Korora	Fish poison, violent gastro intestinal irritant
27.	Cinnamomum camphora	Karpura	Natural camphor, oil, antiseptic.
28.	Cochlospermum nossypium	Kanta palasa	Cigar paste, calieo printings, leather dressing, ice cream industry.
29.	Dalbergia latifolia	Sisso	Furniture, calico printing, tanning, fodder.
30.	Dalbergia sissoo	Sisso	Furniture, railway sleeper, musical instrument, charcoal making.
31.	Dillenia pentagyna	Rai	Cup boards, paneling, tanning and fibre, green manure.
32.	Diospyros embryopteris	Mahar	Kendu tanning, dysentery, diarrhoea.
33.	Diospyros melanoxylon	Kendu	Urapping bidis, dieuratic, laxative, urinary, skin trouble.
34.	Eeblica officinali	Aonla	Lexative, rich source of vitamin, hail dyes, detergent, oil.
35.	Euonymus glaber	Tana	Boxwood, carving and turnery.
36.	Ficus bengalenisis	Bara	Fodder, diarrhoea, rheumatism, tonic, astringent, dysentery, diabetes.
37.	Ficus glomerata	Dimuri	Fodder, diarrhoes, diahotes, piles diarrhoea, match box.

Table 4.2: (Contd....)

Sl.No.	Botanical Name	Oriya Name	Uses
38.	Ficus religiosa	Asastha	Match box, tanning, ulcers, skin trouble.
39.	Firreana colorata	Pani kodal	Ropes, paper pulp, fodder
40.	Garuga pinnata	Kathakusuma	Furniture, cabinetwork, pencils, fodder, brown wrapping paper.
41.	Glochidion lancesolarium	Kalichua	House building, stomach ointment
42.	Gmeling arborea	Gamhari	Furniture, musical instrument, paper making, fever, match box.
43.	Grewia tiliaefoling	Kultho	Fruits edible, poles, frames, tool handle, stamps.
44.	Holoptelea integrifolia	Dauranja	Indoor building, plywood, match boxes, fatty oil.
45.	Hymenoictyon excelum	Barkunda	Teaboxes, jute milling, toys, fodder.
46.	Kydia caalysina	Bana kopasia	Agricultural implements, match boxes, plywood, shuttles.
47.	Lagerstroemia parviflora	Sidha	Building, agriculture, telegraph pool, decorative
48.	Lannea coromandelica	Jia	Furniture, wheel spokes, plywood, Jutemills, railway sleepers.
49.	Litsea polyantha	Bastura	House building, teabox, xandle manufacture, ointment for rheumatism.
50.	Macaranna denticulate	Fera	Fencing, construction of temple hut, tea box, tanning.
51.	Macamna poltata	Tabhari	Sizing paper, tanning matches and paper pulp

(Contd....)

Table 4.2: (Contd....)

Sl.No.	Botanical Name	Oriya Name	Uses
52.	Mallotus philippinensis	Kamala gundi	Dye, Kamala powder, antioxidant for ghee, paints, ointments, match box.
53.	Megnifera indica	Amba	Vit.C, diuretic, uterine haemorrhage, asthma
54.	Mesua ferrea	Nageswar	Boat building, railway sleeper, tool handle, musical instrument, soap making, rheumatism, cough.
55.	Michelia champaca	Champa	Furniture, decorative, fitting, ship building, yellow dye, reneal disease.
56.	Miliusa velutina	Gandha palasa	Building, flooring, furniture, agriculture, match box, purgative.
57.	Mitragvana parviflora	Oudikoim	Turnery, match box, calico printing, colic, tanning.
58.	Nieromelum interinu		Tuberculosis
59.	Ougeinia dalbergioidea	Banjan	Sugarcane, crushers, furniture, cabinet, dysentery.
60.	Oding wodier	Moi	glycoside rhinanthin
61.	Phoenix sylvestris	Khajuri	Vit.C, Mats, nets, toothache.
62.	Phoeba lanceolata		Planking and canoes, fodder, sores
63.	Polyalthia cerasoides	Potmassu	House construction, packing cases, boat building, tumey.
64.	Pongamia glabra	Karanja	Oil, dressing leathers, washing soaps, candles, scabies leprosy, cleaning teeth.

Table 4.2: (Contd....)

Sl.No.	*Botanical Name*	*Oriya Name*	*Uses*
65.	Protium serretum	Rimili	Growing aghagi crop
66.	Pterocarous marsupium	Piasala	Building purpose, railway carriage, wagons, boats, mathematical instruments, dye, oil,paper industry, diabetics
67.	Santalumn album	Chandana	Sandalwood, oil, diuretic, diaphoretic, expectorant sandal tren is parasitic.
68.	Saraca indica	Ashoka	Utrerinine infection, amenorrhoea, internal bleeding tonic, dysenteric, diabetics.
69.	Schleichera olesosa	Kusuma	Oil, sugarmills, agril. implement, wagon building, fuel making, charcoal, rheumatism, soap
70.	Schleichera oleosa	Kusuma	Oil, sugarmills, agril. implement wagon building, fuel making, charcoal, rheumatism, soap.
71.	Semecarous anacardium	Balallia (Cashew)	Source of phenol, light furniture, packing box, leprosy, asthma.
72.	Shorea robusta	Sala	Sleeperwood, agricultural implements, paints, shoe polishes, oil, bidis.
73.	Sideroxylan tomentosa	Kanta bobhula	Anticholeric, oil
74.	Soymida febrifuna	Rohini	Fever, dysentery, vaginal infection, ropes
75.	Spondias mangifera	Ambada	Beats, plywood, diarrhoea, dysentery, rheumatism, regulating menstruation.

(Contd....)

Table 4.2: (Contd.)

Sl.No.	*Botanical Name*	*Oriya Name*	*Uses*
76.	Sterculia urens	Genduli	Gum, matchbox, oil, soap, facilitate delivery, adhesive in plywood.
77.	Streospermum suaveolens	Baropatuli	Constructional work, wagons, furniture, cabinet work, tool hand, firewood, anti cancer fodder.
78.	Strvchnos nuxvomica	Kochila	Nervous disorder, colic, wounds, ulcer, epilepsy, dysentery, fevers, cholera.
79.	Syzygium cumini	Ludham	Dyes, check abortion, fodder, oil
80.	Syzygium cumini	Jamu	Honey, fodder, diabetics, bronchitis, rheumatism
81.	Lamarindus indica	Tontuli	Polishing and cleaning metal things, industries, flower for honey.
82.	Toctona grandis	Saguan	Doors, window frames, furniture making, wagon, railway carriage.
83.	Teorminalia arjuo	Arjun	Railway, boat, house building.
84.	T Belerica	Bahada	Coffee boxes, oil, detergent.
85.	T Chebula	Harida	Ayurvedic medicine, asthma, cardio tonic, laxative gum.
86.	T tomentosa	Asana	Furtinure, window, door, tanning, dying, cardio tonic.
87.	Toona ciliata	Toon	Cigareasens, tea boxes, plywood, ulcer disease.
88.	Tiewis nudiflora	Panigambhari	Matchbox, picture frame, gout, rheumatism
89.	Wrightig tomentosa	Sanchi kurchi	Carving, turney, textiles.
90.	Xylia Xyocarpa	Kangra	Boat, ships, house construction, paper pulp, gonorrhoea and diarrhoea, stop vomiting, manure.

for these natural drugs should be extensively worked out to meet the need of the people.

Illicit cutting of timber trees and transporting the logs by lorries, stealthily during nights and thus developing a considerable commercial racket, also cause considerable demage to the forest (besides reduction of revenue to the state) as many immature timber trees are also destroyed or removed by such illicit methods.

Keeping in view of planned approach for the betterment of rural and tribal population to being with various useful plants collected and studies from the block for example, the Daringbadi is classcified (fig. 4.1) in to (1) Medicinal plants (Table 4.1), (2) Economic importance plants (Table 4.2), (3) Horticultural plants (Table 4.3). Energy plantation species (Table 4.4). Fodder grass species and the various species under these sections are presented under respective tables at the end for purposes of brevity. Such data reveals the extent of plant diversity even in one district. Besides these, the Eastern Ghats from North to South harbour interesting special type of species of their own which are considered as Endemic to the Eastern Ghats and they are of considerable value both Botanically and for National and International exchange programmes for the endemic species of other areas of India and other countries outside India. Such data highlights the standard of biodiversity enen along Dry Deciduous vegetation of the Eastern Ghats. After good study by different workers along the Eastern Ghats, a list of endemic species have been worked out. Out of about 2000 species of flowering plants known so far from the Eastern Ghats, as many as 72 species and 7 varieties (67 Dicots and 9 Monocots under 58 genera and 25 Families) and one Gymnosperm (Cycas beddomei) have been reported so far. Many of such endemic species have been located from Ganjam-Koraput range, Sapparla-Maredumalli area, Nallamala hills, Palakonda range and isolated Sandoor hills. Now many of such earlier dense forest groves are fast getting depleted, thus destroying quite many valuable species. Such

Endemic species along with other useful species of Trees, Shrubs, Climbers and Herbs can also be introduced and grown in Botanical Gardens atleast even a small one for each district for purposes of education, research, public awareness and also for the most important purpose, namely ex situ conservation.

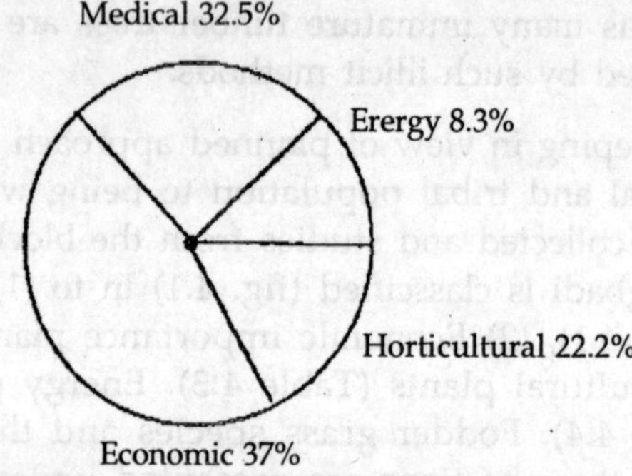

Fig. 4.1

Model Shows Percentage of Medical, Economic, Aorticuttend and Fuel sps of Khond & rd.

Table 4.3: Plants of Horticultural Interest Daringbadi Forests of Kondhmal District

1. Agansome dichotomy K. Schum.
2. Argyeria nervosa (Burm.f.) Boj.
3. Baleria prionitis Linn.
4. Baleria strigosa Wild.
5. Barringtonia acutangula (Linn.) Gaertn.
6. Callicarpa arborea Roxb.
7. Calycopteris floribunda (Roxb.) Lamk.
8. Caralluma adscendens (Roxb.) R.Br.
9. Cassia fistula Linn.
10. Cestrum nocturnum Linn.
11. Clematis smilacifolia Wall.
12. Cleome chelidonii Linn.f.

(Contd....)

13. Clerodendrum inerme (Linn.) Gaertn.
14. Clerodendrum serratum (Linn.) Moon
15. Clerodendrum viscosum Vent.
16. Crossandra infundibuliformis Nees
17. Dracaena temifolia Roxb.
18. Duranta repens Linn.
19. Elytraria acaulis (Linn.f.) Lindau
20. Erycibe paniculata Roxb.
21. Erythoxylum monogynum Roxb.
22. Garcinia xanthochymus Hook.f.
23. Geodorum densiflourm (Lamk.) Sehle.
24. Gmelina arborea Roxb.
25. Gloriosa superba Linn.

Table 4.4: Energy Plantations Fuel Wood Species Daringbadi Forest of Khondhmal District

1. Acacia auriculiformis
2. Acacia nilotica
3. Acacia Senegal
4. Albizia lebbek
5. Alnus nepalensis
6. Azadirachta indica
7. Canthium dicoccum
8. Cassia siamea
9. Cusuarina equisetifolia
10. Derris indica
11. Grevillea robusta
12. Guazuma tomentosa
13. Leucaena leucocephala
14. Memecylon umbellatum

(Contd....)

15. Morinda tinctoria
16. Pithecellobium dulce
17. Prosopis chilensis
18. Prosopis ceneraria
19. Sesbania grandiflora
20. Tamarix aphylla

Forest Products and their Utilisation

Legally all collections are banned from the core area, but they are collected from buffer and periphery of Kondhamala.

Major Forest Products

(i) Timber *(ii)* Firewood

Timber: In the past, Sal was the most important commercial timber but it was banned from 1987. The other important are: (1) Piasal; (2) Dhaura; (3) Gambhari; (4) Sisoo; (5) Koim; (6) Simili; (7) Panjan; (8) Chural; (9) Bankira; (10) Gadikoim; (11) Arjuna; (12) Kusum; (13) Asan; (14) Sidha; (15) Karda; (16) Mahula; (17) Champa.

Firewood: In the past, firewood markets were mainly Berhampur, Bhubaneswar and Cuttack. Now there is restriction for free cutting in Daringbadi reserve forest.

Minor Plant Products: The minor plant products of Kondhamala are Mahua Oil seeds like Mahua (Plate 39), Sal, Castow, Neem, Karanj and Lin, fruits like Harida, Bahada, Amala, Kendu (Plate 40) and Sal leaves, Siali, and Sabai Grass, Rasin and Gum honey and wax, lac and tassar and bamboo.

Forest Based Industry: A number of small scale and cottage industries on forest products have developed in vicinity of Kondhamala. They provide a part of whole time employment to the villagers. Some small scale industries have developed in the periphery of Kondhamala area. Development of these industries has been adequately emphasised in the industrical polity of Orissa (1992) with a view to meeting the change of State growing poverty, unemployment especially in backward and tribal areas. The small scale industries registered with the District Industries Officer, are 13 rice mills, 12 rice-cum flour mills, 12 Oils mills, 8 carpentry units, 3 leather units, 3 rice-cum oil mills, 2 tannery 2 tin container repairing units, 5 sawmills, 2 silver ornament manufacturing units, 4 break making unit, 2 medicinal units and 8 tailoring units.

Oil seeds constitute an important group of commercial forest products. Oil extracted from oil seeds forms an essential fat content on the diet of Indian people. It is also used in medicines, perfumery, varnishes, lubricants, candles, soaps and many other products. Oil cake, the residue after extraction of oil is an important cattle feed, substantially quantity of which are exported. Small oil mills are present at Daringbadi which are producing the oil.

Tassar Cultivation: Tassar rearing is very common in the plain reserves and protected forests of Tumudibandha and Kotagarh and some places and Simanbadi ranges. The cocoons of these areas have a reputation for high yield of Tassar. The control of export and collection of tassar cocoons is under sericulture department. It generates huge employment for cultivation, processing and marketing.

Lac Cultivation: The lac cultivation is made in the plants of Kusum, Palas and Barkoli, now it is declined.

Manufacture of Ropes: Rope making from sabaigrass has developed into a very profitable and important cottage industry. Sabairope is in great demand by the paper mills for tying bamboo bundles.

It is widely used in manufacture of asbestos; sabai ropes are transported to different parts of India from Kondhamala. It is good source of income of the village people.

Siali Fibra: Some tribals living inside the buffer area collected fibre from Siali barkwhich is used for typing cattle and other domestic animals. Sabar Mankadia, Kondha and Pano tribes prepare one type of containers from which is used in extraction of oil from oil seeds by the village people.

Bamboo is another commercial plant of Kondhamala tribes prepare mats, basket hand made fans and umbrella and sell in the local markets. It is the only source of their income. The major commercial importance of bamboo is of manufacture of papers.

Harida, Bahada and Amala in the name of Triphala have a good medicinal value.

Honey is collected by Sabara, Mankadia, Kondh, and Pano tribes from the Kondhamala forest. It has a highly medicinal value in dysentery, cough and diarrhoea, Now-a-days honey cultivation in the village area is financed by the State Government.

Lac, Resine and Gums: These are unique natural products having diversified properties used for many purposes. These are used in manufacture of French polish, gramophone records, micamite grinding wheels, adhesives, cements, wood furning and printing work.

Rock-based Cottage Industries: Leaving aside cultivation some tribes carve statues in Muguni stones. They also prepare pot, plates and thoughts from the black rock, Khiching temple is an example of fine work in Muguni stone.

Animal Products: Products of wild animals have been much sought after from time immemorial. Articles like ivory, peacock feathers, precious skin and musk etc. are highly valued in National and International markets. So tribes of this area are fascinated to collect these articles and sell them through different channels.

REFERENCES

Anonymous (1976): The Wealth of India—Raw Material, Vol. II, C.S.I.R. New Delhi.

Arulomozhian, R.S. Selvaraj, R.; Vijaylakshmi, J; Dhanjayan and Jehangir, K.S. (1992): Three Acaciar Promise Relief for Fuel Wood Shortages, Aqroforestry, Today 4:14.

Barrows, E.M. (1980): Robbing of Exotic Plants by Introduced Carpenter Bees and Honey Bees in Hawaii with Comparative Notes. Bio-tropic 12:23-29.

Fyson, P.F. (1915): Flora of the Nilgiri and Pulney Hill Tops, Vol. 1, (repn. ed. 1974). The Wealth of India-Raw Material, Vol. II. C.S.I.R, New Delhi.

Gamble, J.S. (1915-1936): Flora of the Presidency of Madras, rol_3

Lakhshimi Narayana, K. and Venkaiah, M., (1998): Bio-diversity in the Scared Groves of the North Coastal District of Andhra Pradesh e.g. Proceedings of the National Seminar on Conscrvation of Eastern Ghats, March 24-26, 1998 EPTRI-ENVIS.

Parvathi, A. and Banumathy, G. (1998): Windling Herbal Wealth of Ootacamund; The Eastern Ghats, Proceedings of National Seminar on Concervation of East Ghats, March 24-26, EPTRI ENVIS.

Environment and Pollution Control *A Look*

Dr. Sudhansu Se. Nayak[♣]
Dr. Anil Kumar Sahu[♠]

Introduction

Pollution is defined as an undesirable change in the physical, chemical and biological characteristics of air, water and soil that may harmfully affect the life or create a potential hazard for living organism (Odum, 1971). Degradation of environment with the introduction of man-made extraneous materials to air, water or land is also called 'pollution'. The extraneous agents which cause pollution are called as 'pollutants'. The pollutants may be material or energy which cause pollution when added to the environment in quantities and at a rate which are detrimental to the life of the organisms. Thus, a 'pollutant' is defined as 'any material (solid, liquid or gas) or energy (heat, sound, radioactivity, etc.) that produces harmful effects, when added to the environment.

Pollutants may be divided into two categories:

1. Persistent pollutants;
2. Non-persistent pollutants;

♣ Lecturer in Commerce, R.N. College, Dura, Berhampur-10, Ganjam, Orissa.
♠ Reader in MBA, Berhampur University, Orissa.

1. *Persistent Pollutants*

These are non-biodegradable pollutants. They remain unchanged and stable for a long period when added to the environment as they do not degrade or break down easily into harmless simpler forms.

Examples—DDT, Polythene, Mercuric salts, etc.

2. *Non-persistent Pollutants*

These are 'bio-degradable' pollutants. They degrade or break down into simpler substances in a short time. When added to the environment, they are acted upon or decomposed by microorganisms to be broken down into simpler nutrients that mix with soil.

Examples—Agricultural wastes, garbage, domestic sewage, etc. Major types of pollutants are:

i. Gases

ii. Metals and their salts

iii. Pesticides and agrochemicals

iv Organic matter

v. Radioactive substances

vi. Drugs and pharmaceuticals Noise

vii. Hear

viii. Noise

Air Pollution

Air is the gaseous mixture of troposphere. It is the source of oxygen for respiration of the living organisms. The constituent gases of air and their per cent concentrations are as follows: Nitrogen-78.09 per cent, Oxygen-20.93 per cent, Argon 0.93 per cent, Carbon Dioxide-0.03 per cent, other gases-0.02 per cent, Solid particles per cent.

The composition or air as stated above almost remains constant. As a result, the living organisms get enough oxygen for their respiration. But this composition of air changes when potentially harmful gases and particulate matter are introduced either by natural processes or by human activities. With the increase of harmful gases and solid particles in the polluted air, the concentration of oxygen decreases.

Air Pollutants

The following are the five major primary air pollutants:

1. Oxides of carbon
2. Sulphur dioxide
3. Oxides of nitrogen
4. Hydrocarbons
5. Aerosols

Sources of Air-Pollution

1. *Automobiles*

The air is mostly polluted by the exhaust gases of automobiles which constitute 60 per cent of the air pollution. The automobiles exhausts contain compounds like carbon monoxide, oxides of nitrogen, hydrocarbons, sulphur dioxide, aldehydes, organic vapours and particulate lead. A hydrocarbon, - benzopyrene is released to the air when fuels burn incompletely.

2. *Industries*

Pollutants like flyash, soot and sulphur dioxide are released from thermal power plants using coal. The sulphuric and phosphoric acid units of fertiliser plants produce oxides of sulphur, particulate matter and fluorine. Plants utilising nitrogen produces ammonia, nitrogen oxides and hydrocarbons. The major pllutants produced by the textile industry are cotton dust, nitrogen oxide, chlorine, naphtha vapour and sulphur dioxide.

Chemical plants and pesticide plants release chlorine gas. Steel plants generate carbon monoxide, carbon dioxide, sulphur dioxide, fluorine and dust.

3. *Burning of Wood and Fossil Fuels*

Smoke and carbon dioxide are produced during burning or wood for cooking in homes, hotels and other public places. Burning of fossil fuels like coal, kerosene, petrol, etc. produce carbon dioxide, carbon monoxide, sulphur dioxide, oxides of nitrogen, hydrocarbons, particulate matter and metallic traces.

4. *Lonic Radiation*

Air and water are contaminated by radioactive dusts generated from nuclear tests which find their way into the food chain. LoniZing radiations emanated from radioactive wastes of nuclear reactors Ionise protoplasm in the living organisms.

5. *Agriculture*

Air pollution is also caused by pesticides and fertilizers used indiscriminately in the agricultural fields. A majority of the pesticides do not degrade into simple forms. As they remain stable in the environment, they enter the food chain. Air is also polluted by spray dusts and granules of pesticides produced during spraying of pesticides over infected plants.

6. *Natural Pollutants*

Pollens and spores produced by seasonal plants cause allergic reactions when reach us carried by air. Breakdown of photosynthetic pigments in algae produce carbon dioxide. Volcanic cruption, forest fires, decomposition of natural organic and inorganic materials produce harmful particles and gases which also pollute air.

Control of Air Pollution

- *(i)* Industries can use smokeless fuels;
- *(ii)* Precipitators, scrubbers and filters can be used by industries to check the production of particulate matter;
- *(iii)* By using smokeless chullahs, solar cookers, and biogas, production of smoke can be controlled;

(iv) Industries can use large chimneys;

(v) Positive crank case ventilation and catalytic converter can be used to reduce automobile emissions;

(vi) To control sulphur dioxide concentration in air, the recommended methods are: (a) use of low sulphur fuels; (b) removal of sulphur from the fuel before use; and (c) use of scrubbers;

(vii) Antipollution devices can be fitted to automobiles to filter the emissions.

(viii) Alternate fuels can be developed fro automobiles that can reduce the pollutants in their exhausts;

(ix) Public awareness regarding hazards of air pollution, deforestation and importance of plantation should be created.

Water Pollution

Water is the universal solvent, the elixir of life and also the medium of life activity. About 85 to 90 per cent of the weight of the living body is water. Water is indispensable in our daily life as we use it for various purposes like cooking, washing clothes, cleaning, drinking, etc.

Two-third of the surface of the earth is covered with water. About 97 per cent of the total water present on earth is salt water and remaining 3 per cent is freshwater. The sources of water are:

(i) sea *(ii)* rain

(iii) groundwater, and *(iv)* surface water

The freshwater present in ponds, ditches, steams, rivers, lakes and below the ground is generally used by us for different purposed. Water is called 'pure water' when it is devoid of micro-organisms and has very low amount of salt. This water is used for drinking, cooking, washing, cleaning, industry and agriculture. But through various human activities, organic and inorganic substances are introduced into freshwater and as a result water becomes polluted. Use of polluted water is hazardous to the health and life of living organisms.

Water Pollutants

The following are the major water pollutants:

(i) Sewage and other organic wastes;

(ii) Animal and human excreta;

(iii) Fertilisers, pesticides and insecticides;

(iv) Industrial wastes;

(v) Oil spills;

(vi) Acid rain;

(vii) Metals (like mercury, lead, cadmium, arsenic, etc);

(viii) Open earth mining;

(ix) Temperature;

(x) Radioactivity;

Sources of Water Pollution

1. *Sewage and other organic wastes*

Domestic sewage, industrial sewage, other organic wastes and excrete or human beings and animals are the major pollutants or water in and around human habitations. They facilitate luxuriant growth of bacteria when dumped into water. These bacteria decompose the waters utilising the oxygen present in water and thus cause depletion of oxygen concentration in water. The oxygen-deficient water becomes unsuitable for the aquatic animals to live in it. This polluted water also contains germs of diseases like typhoid, cholera, diarrhoea, jaundice and hepatitis.

2. *Agrochemicals*

The fertilisers, pesticides and insecticides frequently used in the crop fields are washed by following water and thus carried by the latter to other water bodies to pollute them. When the water polluted by agrochemicals is used, causes health hazards for live stock, wild animals, aquatic animals and human being. A good number of pesticides do not degrade easily and remain, stable in the environment and get incorporated in the food chain. They enter the bodies of the organism through food and water and get deposited in the tissues of human beings and animals. The pesticides not only destroy the harmful insects and other pests but also beneficial organisms like earthworms and fishes.

3. *Thermal Pollution*

The rise in temperature of water bodies is detrimental to the life of aquatic animals. The water used for cooling purposes in nuclear reactors, electricity generation plants, steel plants and oil refineries becomes hot and the heated water is generally discharged into ponds, lakes, rivers or seas. As a consequence, the temperature of these water bodies rise, thereby depleting the oxygen concentrations in water. High temperature and low oxygen concentration make the water inhospitable for the aquatic animals.

4. *Oil Spills*

The oil which spills out during transportation on sea or during offshore drilling pollutes the water. Oil being lighter than water, covers the surface of water and thus prevents water from contact with air resulting in graded depletion of dissolved oxygen. The low oxygen concentration of oil covered water affects lives of aquatic animal. Apart from this, when aquatic birds and other animals come in contact with spilled oil, their organs and organ systems become paralysed.

5. *Acid Rain*

Nitrogen oxides and sulphur dioxide produced from natural processes, factories, combustion of fuels and automobile exhausts move in the air, react with moisture and produce sulphuric acid and nitric acid. When rain eater brings these acids formed in the air down to fall in the water bodies, water becomes polluted. They change the pH (hydrogen ion concentration) of water affecting the lives of aquatic organisms.

6. *Metals*

The industrial wastes, discharge dusts, fumes, sludges, waste water, etc. contain metals like mercury, lead, cadmium, arsenic, etc. These metals when released to the environment find their way to the body of human beings and other organisms through food or water and cause damages and disorders in the bodies or the organisms. The metals get increasingly deposited in bodies of organisms at higher tropics levels and the phenomenon is called biomagnifications.

7. *Radioactivity*

Radioactive wastes discharged by the nuclear plants pollute the seas. Radioactive substances of these wastes also spread elsewhere from seas and can cause cancer and damage to the genetic material of the living organisms.

Effects of Water Pollution

1. *Eutrophication*

The domestic sewage and run-off water from agricultural fields release inorganic and fertilisers into different water bodies. As a result, the nutrient load of these water bodies gradually increase which facilitates algal bloom in them. Water bodies with algal bloom appear totally green. Algal bloom and other organisms present in these water bodies promote a large population of decomposers like bacteria and fungi. When the algae and other organisms die, they are decomposed by these decomposers. The dissolved oxygen concentration of these water bodies and thus create problems for the fishes. The phenomenon of depletion of concentration of water as a consequence of increase in nutrient load is called eutrophication.

2. *Ecological Magnification*

Polluted water contains many unnatural toxic substances like metals and pesticides which remain stable in the environment without being degraded. These substances find their way into the body of living organisms through food chain. The body of the organisms neither can degrade them into harmless simple components nor can excrete them easily. As a result, gradually they accumulate in the tissues of plants and animal and pass from one trophic level to the other in the food chain to reach main finally. The concentration of these chemicals or substances is very high in the organisms of higher trophic levels of the food chain. Increasing accumulation of the stable toxic substances in the tissues of organisms at different trophic levels of food chain is referred to as 'ecological magnification'.

3. *Effect of Industrial Discharges*

A variety of harmful substances like acids, alkalis, dyes and many other toxic chemicals are present in the industrial effluents which are released to the different water bodies. The addition of these harmful substances changes the pH of water and effects adversely the distribution and activities of aquatic plants and animals. Again stress factors produced by substances like acids, alkalis and detergents present in the industrial discharges cause the death of fishes.

4. *Effects of Agrochemicals*

There are many pesticides which are not biodegradable and once introduced, remain stable in the environment for a very long time. These agrochemicals find their way into the bodies of the organisms through food chain and get magnified in higher trophic levels. The stable agrochemicals, cause various disorders in the bodies of living organisms including human beings. Organochlroine pesticides like DDT cause damages in brain and liver and organophosphorus insecticides cause sweating, abdominal cramp, chest discomfort, vomiting, headache, nervousness, etc.

5. *Effects of Metals*

Magnification of heavy metals released to the environment through industrial wastes occurs in every stage of transfer in the food chain. Metals like mercury, lead, cadmium, arsenic, nickel and chromium are highly toxic in the body of human beings. The disorders caused by these metals in humans are as follows:

6. *Red Tide*

When costal waters become enriched with nutrients of the discharged sewage, blooms of dinoflagellate occur in them. These dinoflagellate produce toxic substances which cause death of marine fauna. The dinoflagellage blooms make the tidal water appear red in colour, hence the name 'red tide'.

Control of Water Pollution

1. Water pollution can be controlled by enacting stringent laws to punish the people who pollute the water.
2. Regular checks and supervisions should be undertaken to maintain the quality of drinking and potable water.
3. Scientific methods should be adopted to handle and dispose the garbage, sewage, excreta and industrial wastes.
4. The industrial discharges should be released to the water bodies only after their neutralisation and proper treatment.
5. Instead of releasing to the water bodies, the effluent water can be, used for growing algae and aquatic plants to produce biomass for use in biogas plants.
6. Large shallow oxidation or stabilising ponds can be constructed to store the domestic sewage and industrial effluents. Green algae and bacteria will gradually grow in such ponds by utilising the nutrients present in them thereby purifying such waters. The water or these pond

can be used for irrigation and contains enough nitrogen, phosphorous and potassium.

7. Public awareness should be created regarding causes and effects of water pollution.
8. The germs of the water bodies including the sewage should be killed time to time using lime and bleaching powder.
9. Method like adsorption, electro dialysis, ion exchange, reverse osmosis, etc. should employed to remove or alter the pollutants like radioactive, chemical and biological substances from water.

Soil Pollution

Soil is the upper layer of earth's crust. Plants grow on soil. Hence the growth of the plants is directly affected by the texture, composition and water content of the soil. Like air and water, soil is also polluted by human activities.

Chief pollutants of the soil are the solid wastes like metals, plastics, polyethylene, human and animal excreta, glass, paper, rubber, building materials like sand, cement, bricks, etc. Many of these solid pollutants are not biodegradable. They not only pollute the surrounding environment by generating foul smell and poisonous gases but also change the texture and composition of soil affecting of the animals and plants inhabiting on it.

Sources of Soil Pollution

1. Domestic wastes
2. Industrial wastes
3. Agricultural wastes
4. Animal wastes
5. Community wastes

Effects of Soil Pollution

In towns and cities, a large amount of solid wastes is dumped on the adjoining soil daily, where soil pollution has become a major problem. As the solid wastes in such thickly populated areas are not collected, disposed or dumped properly, the soil of those human habitats becomes highly polluted and caused a lot of problems as follows:

(i) Breeding of disease carriers like flies and mosquitoes can take place on the decomposed organic solid wastes;

(ii) Organic matters may breed microbes;

(iii) Decomposed organic matters produce foul smell;

(iv) Solid wastes can be washed away by run-off water during rainy season and pollute other water bodies;

(v) Non-biodegradable solid wastes like polythene, plastic, glass, metal, etc. when dumped on soil create problems for plants.

Control of Soil Pollution

Soil pollution can be controlled by the following three ways:

(i) Throw-away output approach

Dumping of solid wastes on land or water bodies or burning is known as throw-away output approach of solid waste disposal. But instead' of protecting the environment, this approach pollutes the place of dumping, or, burning.

(ii) In-put approach

By reducing the production of solid wastes, we can prevent soil pollution caused by their harmful effects.

(iii) Resource recovery output approach

Recycling of solid wastes can be undertaken to convert them into many useful products. Thus, they can be a resource rather than a liability. The followings are some of the examples of this resource output approach.

(a) Packing card boards, hand-made drawing sheets, newsprint, unbreakable dolls and many other

durable items can be produced from recycled waste papers.

(*b*) Many reusable items can also be produced from recycled solid wastes like glasses, plastics, polythene, etc.

(*c*) Compost can be prepared from organic wastes by dumping them under soil. This compost is a rich source of nutrients for plants when used as manure.

(*d*) Biogas can be produced form excreta of man and other animals for use as fuel and for lighting houses.

(*e*) Ash is a solid waste produced form thermal power plants run on coal. This hazardous environmental pollutant can be used to make bricks and concrete and as manure in agricultural fields.

Noise Pollution

Noise can be defined as any sound that is unwanted and unpleasant and causes discomfort and annoyance. Both health and mind of an individual get effected by noise. Sound is measured by a unit called decibel. Sound becomes injurious to our ear when it is of more than 130 decibels. Again, sound is unwanted when it crosses 100 decibel mark.

Sources of Sound

(*a*) Industries or factories

(*b*) Traffic

(*c*) Noise of communities and gatherings

(*d*) Electrical or electronic gadgets.

(a) Industries

Machineries of industries run in compact spaces produce unpleasant sound of higher decibels. Such sounds are injurious to the health and mind of workers.

(b) Traffic

Number of automobiles on the roads has been increasing day by day. The vehicles like motor cycles, scooters, cars, buses, trucks, auto rickshaws, aeroplanes, etc. not only pollute air by producing harmful gases in their exhausts but also produce irritating, unpleasant and unbearable sound. These sound can cause physical and psychological illness, especially in people of urban areas adjacent to major roads.

(c) Communities and gatherings

Gatherings of people in festivals, processions, marriages, meetings, rallies, etc. is another major sources of noise. The sounds produced from chanting slogans, high pitched music, speeches, songs, use of microphones, etc. associated with such gatherings are also injurious to our health.

(d) Electrical and electronic gadgets

Frequent use of electrical and electronic gadgets like radio; TV, telephone, grinder mixer, washing machine, cookers and many such appliances also becomes a major source of unpleasant sound.

Effects of Noise Pollution

Noise is always irritating, annoying, unpleasant, unbearable and injurious to our health. It affects the work of people of places like hospitals, schools, offices, libraries, etc. which require a calm, peaceful and serene atmosphere. The followings are some of the important effects of sound or noise pollution:

(i) Irritation, annoyance

(ii) Reduction of work efficiency

(iii) Impairment of loss of hearing

(iv) Insomnia or sleeplessness

(v) Hypertension (or increase in blood pressure)

(vi) Nervous disorders and brain damages

(vii) Sweating, nausea and fatigue

(viii) Ailments of stomach and brain

(ix) Increase in cholesterol level in blood leading to hypertension, heart ailments and strokes.

Control of Noise Pollution

Control of noise production at the source

(i) By locating the chief sources of sound production i.e. industries, aerodromes, highways, etc. away from human settlements.

(ii) By discouraging or preventing or banning establishment of human habitations close to industries, aerodromes, highways, etc.

(iii) By making the use of silencing devices in automobiles, aero planes, industries and other such sound producing machines or electronic and electrical gadgets mandatory.

(iv) By planting trees on both sides of highways, streets, factory premises and near airports and at different places of towns and cities since plants can absorb noise efficiently.

2. *Control of transmission of noise*

If the sound production areas are covered and enclosed by soundproof tiles and high rise walls respectively, the transmission of irritable sound can be prevented or reduced. This purpose can also be served if thick plantat:on can be undertaken around the factories or industries.

3. *Control by protecting the victims of sound pollution*

Device like earplugs and earmuffs should be provided to people who have constant and continuous exposure to harmful irritable proof and all such places of work sound-proof.

4. *Control by enacting laws*

Laws should be enacted to ban or control production of irritable sound from loud speakers, crackers, processions, transportations, machines, etc. during sleeping hours especially in the places where hospitals, schools, colleges, libraries, etc. are located. Adequate laws should also be enacted for banning high pitch horns in the vehicles without silencers and for limiting volume or radios, music systems, TVs while listening.

REFERENCES

1. *A Text book of Environmental Education*, Goyal Brothers Prakasan, New Delhi, 2006.

2. Patnaik, Dr. S.C. & Routray, Dr. P.K. *Environmental Education*, Modern Publishers, Jalandhar, 2006.

3. Das, B.C., *Environmental Education*, Kalyani Publishers, Cuttack, 2006.

4. Satapathy & Das, *Environmental Education*, Kalyani Publishers, Cuttack, 2006.

5. *Comprehensive Environmental Education*, Laxmi Publications (P) Ltd., New Delhi, 2006.

6. *Environmental Education*, Frank Brothers and Company Publishers Ltd, New Delhi, 2006.

7. Sahu Dr. B.K., Mohapatra, P.K. & Dash, Dr. N., *Elements of Environmental Education*. Subnam Pustak Mahal, Cuttack, 2006.

8. Panigrahy, Dr. A.K. & Sahu, Dr. Alka, *Environment Science*, 2005.

Water Crisis—Needs a Holistic Approach

*Bhagaban Padhy**

Everything is originated from water and everything is sustained by it. Even everything will be perished due to lack of water. Water according to eminent Greek philosopher Pindar "is the best of all things". At present only 0.26 per cent of the world's water is accessible for use. The rest 97.5 per cent of it is salt water contained in the ocean and 2.24 per cent is in polar ice caps, glaciers and deep grounds. However, the demand for water is continuously expanding and is expected to increase nine fold, more than triple the rate of population growth by the year 2025. As a result the number of water scarce countries is expected to rise upto 35 by the year 2020; out of which 20 will be absolutely in the grip of water scarcity. There are a number of reasons responsible for water crisis. The most important are global warming, huge urbanisation and industrialisation of the world economy. All these have led to severe threat for the living organisms on the earth. Thus, proper management of water resource in a sustainable manner is necessary in order to meet the basic requirement of all the citizens right now and in future.

* Lecturer in Economics, A.N. College, Dharakote, Gm. (Orissa), E-mail: bhagaban_padhi@sify.com

Development of water resource is particularly important for India in view of its arid monsoon climate. No doubt, since independence great emphasis has been laid on water resources development with almost 25 per cent of the current budget being invested for the said purpose. But the alarming trend of population growth, urbanisation and industrialisation aggravates the situation, which is beyond imagination of planners. Therefore India needs proper management of water resource in order to meet the basic requirement of all the citizens right now and in future. That is why the already available scarce but precious resource should be planned developed, conserved and managed on an integrated and environmentally sound basis keeping in view the socio-economic aspects and growing needs of the citizens of 21st century. Besides, efforts, to develop, conserve, utilise and manage this important resource in a sustainable manner have to be guided by the national perspective.

Uses of Water

Water determines the mode and level of development of the country, for it is one of the most important components of economic development. And it is probably the most important resource in the world since it is required for:

- Sustenance of human life—Physiological process, sanitation
- Agriculture activities—metabolic process for food production, evaporation cooling of earth.
- Industrial needs—heat transfer process, chemical reactions and nuclear fusion power.

Sector-wise Utilisation of Total Water Resources in India

The table given below shows the sector-wise utilisation of total water resources in India at present.

Table 6.1: Sector-wise of Total Water Resources

Purpose	*Present utilisation (bcm)*	*Percentage*
Domestic	30	4.9
Industries	20	3.3
Irrigation	501	82.9
Power generation	20	3.3
Others	34	5.6
Total	605	100.00

Source: National Water Development Agency.

Future Demand for Water

Table 6.2: Shows the Future Demand for Water

Purpose	*Year 2010 (bcm)*	*Year 2025 (bcm)*
Irrigation	688	910
Domestic	56	73
Industrial	12	23
Energy	5	15
Others	72	72
Total	813	1093

Source: National Water Development Agency.

Declining Water Resources

The annual per capita availability of fresh water in 1951 was 5,177 cubic meters, that declined to 1,869 cubic meters in 2001 and is likely to fall further to 1,341 and 1,140 cubic meters in 2025 and 2050 respectively. It is generally presumed that if per capita level falls to 1,000 cubic meters, it could seriously affect the health and economic activity of the entire country. This will affect 25 per cent of India's geographical area and 21 per cent of total population.

Table 6.3:

State	*Water requirement in 2020*			*Water availability*			*Ratio of requirements to availability*
	Agriculture	*Non.Agri*	*Total*	*Surface water*	*Ground water*	*Total*	
Andhra Pradesh	84.6	9.8	94.4	65.9	35.2	101.1	0.93
Assam	3.9	1.7	5.6	10.4	24.7	35.1	0.16
Bihar	48.6	7.8	56.4	18.3	33.5	51.8	1.09
Gujarat	3.8	15.4	79.2	22.1	20.4	42.5	1.86
Haryana	34	4.8	38.8	28.6	8.5	37.1	1.05
Himachal Pradesh	0.7	0.6	1.3	NA	0.4	NA	NA
Karnataka	35.7	6.3	42	51.5	16.1	67.6	0.62
Kerala	5.5	3.9	9.4	17.7	7.9	25.6	0.37
Madhya Pradesh	51.1	10.3	61.4	118.2	50.9	169.1	0.36
Maharashtra	73.3	25.7	99	18.3	37.9	56.2	1.76
Orissa	27	3.5	30.5	25.7	20	45.7	0.67
Punjab	38.0	4.3	42.9	87.4	18.7	106.1	0.40
Rajasthan	25.8	7.3	33.1	NA	12.7	NA	NA
Tamil Nadu	46.7	10.7	57.4	44.9	26.4	71.3	0.81
Uttar Pradesh	123	20.2	143.2	178.5	83.8	262.3	0.55
West Bengal	15.2	7.8	23	14.4	23.1	37.5	0.61
All India	677.3	147.4	824.7	690.3	420.2	1110.5	0.74

Source: Based on Estimate of availability and requirements presented in Chopra Goldar.

Table 6.4: Water Availability and Water Balance in Various River Basins in India

		Water availability per capita per annum (cubic meters) 2025	2050	*Water Balance* 2050
1	Indus	1005.7	836.9	Surplus
2a	Ganga	705.5	622.5	
2b	Brahmaputra	1174.8	1311.7	
2c	Barak	857.1	866.7	
3	Godavari	1324.4	1187.2	Surplus
4	Krishna	781.4	619.4	Surplus
5	Cauvery	609.7	497.5	
6	Subarnarekha	570.5	518.3	
7	Brahmani-Baitarini	1605.8	1848.7	Surplus
8	Mahanadi	1607.4	1553.7	Surplus
9	Pennar	780.8	726.1	Surplus
10	Mahi	587.9	538.9	
11	Sabarmati	91.3	62.5	Deficit
12	Narmada	821.6	622.5	
13	Tapi	821.6	622.5	
14	West flowing rivers between Tapi and Kanyakumari	493.8	305.4	
15	West flowing between Tandri and Kanyakumari	627.9	515.6	NA
16.	East flowing rivers between Mahanadi and Pennar	799.5	704.7	Surplus
17.	East flowing rivers between Pennar and Kanyakumari	198.8	144.3	Deficit
18.	West flowing rivers of Kutch, Saurashtra and Luni	407.6	340.9	Deficit
19	Areas of inland Drainage in Rajasthan	NA	NA	
20	Minor rivers draining into Bangladesh and Myanmar	4666.7	3733.3	NA
	Total	779.3	660.0	

Source: Based on (I) Water Resources Development Plan of India . Policy and issues.

Future Water Requirements

The natural water commission has estimated that the total water requirements of the country in the year 2050 would be about 973 bcm on the lower slide and 1180 bcm on the upper side depending upon the actual population growth. However the Ministry of Water Resources on the basis of other studies has estimated India's water requirement to be around 1093 bcm for the year 2025 and about 1,447 bcm for the year 2050.

However, most natural waters are not suitable for human consumption and in course of time they are becoming scarce and unsafe for human consumption. It is estimated that only 0.10 per cent of total water resources available on the globe are fit for drinking purposes. Hence, men must cease to regard water resources as an inexhaustible gift of nature and use it with greatest care and proper attention. Besides men should attempt unitedly to conserve the same.

It is evident from the above tables that the globe is heading towards increase in the magnitude of water uses and scarcity.

In this backdrop attempts have been made in this chapter to discuss the following dimensions in relation to water in five sections.

Section 1 - Factors Responsible for Water Scarcity.

Section 2 - Water Pollutants and their Impact on Health.

Section 3 - National Water Policy for Sustainable Management.

Section 4 - Programmes, Plans and Projects.

Section 5 - Suggestion.

Section 1

Factors Responsible for Water Scarcity

The following factors are responsible for shortage of water:

- Failure to provide equalisation of surface water run off;
- Over concentration of population and industries;
- Withdrawal of water from underground aquifers in excess of natural recharging rate;
- Water wastage since it is cheap;
- Failure to use salt water in place of fresh water because it requires expensive corrosion resistant materials;
- Faulty selection of industrial sites;
- Lack of industrial co-ordination;
- Faulty process of consumption.

Section 2

Water Pollutants and their Impact on Health

The following are the water pollutants, which cause a great loss to living orgañisms in different ways.

1. *Sewage and Other Organic Wastes*

Domestic sewage, industrial sewage, other organic wastes and execreta of human beings and animals are the major pollutants of water in and around human habitations. They facilitate luxuriant growth of bacteria when dumped into water. These bacteria decomposes the wastes utilising the oxygen present in water and thus cause depletion of oxygen concentration in water. The oxygen deficient water becomes unsuitable for the aquatic animals to live in. This polluted water also contains germs of diseases like typhoid, cholera, diarrhoea, jaundice and hepatitis.

2. *Inorganic Chemicals*

Inorganic chemicals include dangerous metals such as mercury, lead, arsenic, sodium, nitrates and nitrites, asbestos, fluoride, chromium etc. Some inorganics are associated with a variety of health problems. Lead is known to cause brain damage and is highly toxic to infants and pregnant women, even in minute doses. It usually comes from lead water pipes, old pipe solder or lead based brass fixtures. Similarly arsenic is known to cause respiratory and skin cancer, nervous disorders. And mercury may cause disorder of central and peripheral nervous system, numbers of lips, limbs, muscles etc.

3. *Agro Chemicals*

The fertilisers, pesticides and insecticides frequently used in the crop fields are washed by flowing water and thus carried by the latter to other water bodies and pollute them. When the water polluted by agrochemicals is used, causes health hazards for livestock, wild animals, aquatic animals and man. A good number of pesticides do not degrade easily and remain stable in the environment and get in corporated in the food chain. They enter the bodies of the organism through food and water and get deposited in the tissues of human beings and animals. These pesticides not only destroy the harmful insects and other pests but also beneficial organisms like earthworm and fishes.

4. *Radons and other Radionuclides*

Radons and other radionuclides can enter into your body through the air you breathe, no need to drink the water. They come from naturally occurring radon, uranium and radioactive waste and are known to increase cancer risk.

5. *Oil Spills*

The oil which spills out pollutes the water. Oil being lighter than water, covers the surface of water and thus prevents water from contact with air resulting in depletion of dissolved oxygen. The low oxygen concentration of oil covered water affects lives of acquatic animals. Apart from this, when acquatic birds and other animals come in contact with spilled oil their organs and organ systems become paralysed.

6. *Acid Rain*

Nitrogen oxides and sulphur dioxide produced from natural processes, factories, combusion of fuels and automobile exhausts move in the air, react with moisture and produce sulphuric acid and nitric acid. When rain water brings these acids formed in the air down to fall on the water bodies, water becomes polluted, they change the pH (hydrogen ion concentration) of water affecting the lives of acquatic organisms.

7. *Radio Activity*

Radio activity wastes released by the nuclear plants pollute the seas. Radio active substances of these wastes also spread elsewhere from seas and can cause cancer and damage to the genetic material of living organisms.

8. *Thermal Pollution*

The rise in temperature of water bodies is detrimental to the life of acquatic animals. The water used for cooling purposes in nuclear reactors, electricity generation plants, steel plants and oil refineries becomes hot and this heated water is generally discharged into ponds, lakes, rivers or seas. As a consequence the temperature of these water bodies rise, thereby depleting the oxygen concentration in water. High temperature and low oxygen concentration make the water inhospitable for the acquatic animals.

Risks to Children

These water contaminants are very harmful for infants and young children because they face risks as discussed below:

1. *Nitrates*

Nitrates and nitrites in the water can interfere with infants ability to absorb oxygen and can lead to "blue-baby" syndrome (methanoglobinemia) which can result in death.

2. *Arsenic*

Arsenic may cause low birth weights, spontaneous abortions and like problems.

3. *THMs*

Trihalomethanes (THMs) are sometimes created after water has left a treatment plant, when chlorine may react with leaves and other decaying animal and plant matter. A substance that can result from this process is chloroform, a suspected carcinogen.

4. *Pesticides*

Malignancies (in children) linked to pesticides may cause leukemia, neuro blastoma, soft tissue sarcoma and cancer of brain etc.

5. *Biological Contaminants*

E. Coli, Giardia and cryptosoporidia cysts can cause gastro intestinal problems where dehydration from diarrhoea moderate levels of BDPs over a life time increases the risk of cancer.

Contamination can also cause a complication called hemolytic uremia syndrome, in which the red blood cells are destroyed and kidneys fail.

Other Effects: Fluoride

Calcium fluoride is useful to the body and not sodium fluoride the toxic waste by product of the aluminum refining industry that is added to some municipal water supplies. Recent research indicates that exposure to excessive amounts of this chemical during pregnancy can cause brain defects in the offspring.

Chlorine

Chlorine is a well known germicide. It is added to water to kill harmful water borne bacteria before they kill us. However chlorine does not kill viruses in water and often contains carbon tetrachloride, a known carcinogen.

Section 3

For the above mentioned reasons—a comprehensive policy on water has become necessary in the face of the growing number of social, economic and environmental issues surrounding water resources in India. Even it was felt in ancient India when steps were taken to conserve and utilise water properly. This is clear from Arthasastra of Kautilya (400 B.C.) and Astadhyayi of Panini (700 B.C.). Construction methods and materials of dam, tanks etc. bank protection, spill ways and other considerations mentioned in the ancient books reflect the high stage of development of water resources and hydrology in ancient India.

At present the management of India's water resources falls under the jurisdiction of a number of Government agencies, although the primary responsibility for the development of water belongs to the individual states.

The Ministry of Water Resources (MOWR) is the principal agency responsible for water in India as such, oversees the planning and development of the resource, from policy formulation to infrastructure support. Other Central Departments working in water are:

(a) Ministry of Agriculture: Watershed development and irrigation.

(b) Ministry of Rural Development : Watershed development and drinking water provision.

(c) Ministry of Industry: Industrial uses of water

(d) Ministry of urban development: Urban drinking water provision and sanitation.

(e) Central Pollution Control Board: Water quality monitoring.

(f) Indian Council of Agriculture Research: Development of water management techniques.

National Water Policy

Growing number of social, economic and environmental issues surrounding water resources in India compelled the Government heavily as a result National Water Resources Council was formed in the year 1987 which adopted National Water Policy (NWP) and the document was submitted to parliament for implementation. The NWP is the primary document stating the position of the Government of India on water resource issues ranging from drought and flood management to drinking water provision. The policy serves as a guideline to help planners and managers to develop the country's water resources to their maximum potential.

Acts and Regulations

Water in India is also governed under three different Acts: The Environmental Protection Act (1986), the River Boards Act (1956) and the Inter State Water Disputes Act (1956).

1. *Environmental Protection Act*

The Act is concerned with the "Protection and improvement of the human environment" and such does not focus solely on water resources issues. However the focus of the environmental protection Act on water is in terms of protecting water from environmental pollution.

2. *The River Boards Act*

The River Boards were designed to advise the Central Government on development opportunities, co-ordinate activities and resolve disputes. The Boards were required to provide advice to the Government on the following topics:

- Prevention of pollution;
- Conservation with a view to control and optional use of water resources;
- Promotion and operation of schemes related to irrigation, water supply and drainage;
- Promotion and control of navigation;
- Other duties as deemed necessary.

3. *The Inter State Water Disputes (ISWD) Act*

Since the majority of rivers are shared between neighbouring states, the ISWD Act was enacted in 1956 to solve disputes related to distribution or control of the rivers.

The Act gives the Government power to constitute Tribunals to intermediate the disputes and till date the following Inter State Water Tribunals have been established:

- Godavari Water Disputes Tribunal (April 1969)
- Krishna Water Disputes Tribunal (April 1969)
- Narmada Water Disputes Tribunal (October 1969)
- Ravi and Beas Water Disputes Tribunal (April 1986)
- Cauvery Water Disputes Tribunal (June 1990)

Section 4

Programmes, Projects and Plans

A large number of irrigation projects have been launched and different programmes and plans made since independence for conservation and development of scarce water resource.

Hydrology Project

It was decided to launch hydrology projects beginning from 1995-96 at an estimated cost of Rs.609.2 cr. For the said purpose the World Bank has extended a credit assistance of US dollar 142 million. The project covered the peninsular states of India namely Andhra Pradesh, Orissa, Maharastra, Gujarat, Madhya Pradesh, Karnataka, Tamilnadu and Kerala. The main objective of the project is to improve the infrastructure, collection and management of hydrometric and water quality data and the use of such data for water resources evaluation and management.

Command Area Development Programme

A centrally sponsored Command Area Programme was launched in 1974-75 with the main objective of improving irrigation potential and optimising agricultural productivity and production from the irrigated areas by integrating all functions related with irrigated agriculture.

Flood Management

In order to protect the flood prone areas by forecasting

on inter state rivers through its field divisions throughout the country.

Groundwater Development

At present more than 70 per cent of the population uses ground water for its domestic needs and more than half of irrigation is provided from this source. Therefore steps were taken to develop the total replenishable groundwater in India.

National Commission for Integrated Water Resources Development Plan

The Ministry of Water Resources has constituted a national commission for integrated water resources development plan. The aims and objectives of the plan are as follows:

- To prepare an integrated water plan for development of water resources for drinking, irrigation, industrial, flood control and other issues;
- To suggest modalities for transfer of surplus water to deflicit area by inter linking rivers with a view to achieve the abovementioned objectives;
- To identify important on going projects as well as new projects which should be completed on priority basis;
- To identify a technological and inter-disciplinary research plan for the water sector with a view to maximize the benefits;
- To suggest for physical and financial resources generation strategies for the water sector and any officer related issue.

The National Water Board

The National Water Board was constituted by the Government in September 1990 to review the progress and implementation of the National Water Policy and report to the National Water Resources Council for taking effective measures for systematic development of the country's water resources.

National Water Resources Council

The National Water Resources Council was established by the Government of India in March 1983 as a National Apex Body for development of country's water resources.

Central Water Commission

The Central Water Commission was constituted for initiating coordinating and implementing in consultation with the state Governments concerned schemes for the control, conservation and utilisation of water resources for the purposes of flood management, irrigation, navigation and water power generation throughout the country.

Central Water and Power Research Station

Central Water and Power Research Station (CWPRS) established in pure provides research and development support to a variety of projects in the areas of water energy resources development.

Central Groundwater Board

Central Groundwater Board (CGWB) was established with a view to carryout scientific development and management of ground water resources from national perspective. The Board through its branch network in the areas having specific problems like water logging, declining ground water levels and ground water pollution etc. with top priority special emphasis is given on monitoring of ground water levels and water quality through a network of about 15,000 observation stations in the country.

Ganga Flood Control Commission

Ganga Flood Control Commission (GFCC) was established in April 1972 and has been mainly assigned the task of preparing comprehensive plans of flood management of river systems in the Ganga sub-division.

Water and Consultancy Services (India) Limited (WAPCOS), a public sector undertaking under the ministry of water resources, provides integrated technical consultancy services in the water and power sectors in India and abroad.

National Projects Construction Corporation Limited (NPCC) a Government of India enterprise, which was established in 1957 tries for development of infrastructure in the core sectors of irrigation and water resources, power and heavy industry.

National Institute of Hydrology (NIH)

National Institute of Hydrology (NIH) was established with head quarters at Rourkela by the Government of India in 1979 for undertaking, promoting and coordinating systematic scientific works in all aspects of hydrology.

National Water Development Agency (NWDA)

The Government of India established National Water Development Agency (NWDA) in 1982 with a view to transfer surplus water available in some regions to water deficit areas by interlinking rivers.

The survey and investigation of six links namely Manas-Sankosh, Tista Ganga link, Sarda-Yamuna link, Ganga-Damdor-Subarna rekha link, Yamuna-Rajasthan link, Chunar-Sone barrage link proposals are under consideration.

Section 5

Suggestions

In spite of the above mentioned policy measures, programmes and plans for sustainable development of water resource proved futile for a number of reasons like financial crisis, lack of interest, mis-management and haphazard decisions. Hence water management has become one of the important issues of the globe today. That is why for successful management and conservation of water and sustainable development of the scarce resources the following measures should be taken at an earliest possible:

1. *Water Literacy Campaigns*

Water literacy campaigns should be conducted to create awareness among the people regarding scarce water resources. People should be educated and motivated. Universities, voluntary organisations, trade unions, student organisations and media have to play important role in bringing mass awareness.

2. *Inter-linking of Rivers*

The project of inter-linking of rivers in India is expected to provide multidimensional benefits. The expected benefit are augmentation of irrigation facilities, potable water supply for rural and urban areas, industrial water supply, more generation of hydropower, inland navigation, ecological upgradation due to minimum flow guarantee in rivers, sizeable employment generation, flood and drought mitigation, increased three forming and many other indirect benefits. Thus it is a programme for greening India.

3. *Rain Water Harvesting*

Steps should be taken to harvest rain water and ground water recharge scheme be introduced in all States and Union Territories.

4. *Recycling and Reuse of Municipal and Industrial Waste Water*

Recycling and reuse of water can meet the domestic and agricultural needs to some extent.

5. *River Water Harvesting*

Excess water of rivers can also be harvested and camolised to other areas for crop production and other uses. But care should be taken not to disturb the underground balance of the basins. In urban areas harvesting of water should be made compulsory so that the water stored could be used for other than drinking purposes to ease pressure on this scarce resource.

6. *Check on Excess Digging of Wells*

Excess digging of wells in severely affected areas should be avoided.

7. ***Flood Management***

Proper management of floods will result in recharging of ground water aquifers. That is why steps should be taken in this regard.

8. ***Restriction on Water Privatisation***

Privatisation of Water leads to wastage of scarce but valuable water resources and monopoly. So government should take necessary steps to check it.

Conclusion

To conclude playing with water means serious threats to the survival of living biotic organisms. The prediction of Sunderlal Bahuguna *"the acute scarcity of water may force the powerful nations to wage a new global war for the control of the depleted sources of water"* may also be true. So we should respect the united nation's message *"water is a scarce resource to be managed and protected"*.

REFERENCES

1. *India Vision 2020*, Academic Foundation, New Delhi, 2003.
2. Kalam Abdul APJ, "Integrated Water Mission, *Yojana*, June 2005.
3. Duggal V.K. "Drinking Water and Sanitation " *Yojana*, June 2005.
4. Ghosh S. and Verma H.N. "Water in 21st Century" *Kurukshetra*, January, 2002.

Problems of Rural Water Supply and Sanitation

*Prof. Manas Bandyopadhyay**

More than 80 per cent of the Indian population live in rural areas but it is a fact that about 90 per cent have some form of water supply with about 20-25 per cent piped water supply. The major problem of covering all the rural population is due to the scattered nature of the villages, inaccessibility and non-availability of other service facilities. So far emphasis has been given on quantity but not quality.

Rural communities are generally small and poor and so the various elements of the scheme must be simple and comparatively cheaper requiring the minimum skilled supervision and maintenance.

Although provision of water supply has generally received priority in national planning it can be over-emphasised that full health benefits from water supply cannot be achieved without proper sanitation.

Types of wells and Tubewells

Wells with tube as the structural material is generally known as tubewells. The water has to be pumped in almost

---

* Department of Civil Engineering, IIT Kharagpur-721302.

all cases except for artesian wells.

Classification of wells

(a) Flow of water—Artesian, sub-artesian water fed wells.

(b) Materials of construction—Masonary wells, Ring wells, Tubewells.

(c) Deep wells and Shallow wells.

(d) Method of construction—Sinking, water jet boring etc.

Aquifers

An aquifer is a water bearing stratum of formation capable of transmitting water in quantities sufficient to permit development. Aquifers may be considered as falling into two categories, confined and unconfined depending on weather or not a water table or free surface exists under atmospheric pressure.

Shallow Well

A shallow well is one which is constructed by tapping the uppermost water bearing stratum.

Deep Well

A deep well is usually constructed by tapping a water bearing stratum underneath an impervious stratum.

Yield of a shallow well is uncertain due to large variation of ground water level. Quality may be deteriorated due to the close existence of dug well latrines and waste dumps.

Quality of Groundwater

Generally ground water is clear and colourless but are harder than surface water due to presence of minerals, calcium and magnesium. Bacterially ground waters are much better than surface water excepting where sub surface pollution exists. Apart from this, groundwater may contain, Iron and

Manganese which causes rust spots of fabrics and utensils. It may also fabrics promote growth of iron bacteria causing taste and odour and imparts colour to water. Excessive fluorides more than 1.5 mg/l may cause dental fluerons and less than 0.6 mg/l may cause dental cancers to children. Nitrates may sometimes cause disease like blue babies or nitrate poisoning to infants if exists in excess of 40 mg/l. Arsenic more than 0.05 mg/l may be toxic to human beings.

Groundwater in coastal aquifer may have the problem of salinity due to the presence of excessive salts.

Maintenance of Wells and Tubewells

In order to get a reliable supply of good quality water the wells and tubewells are to be kept under continuous watch and care.

The following steps are to be taken:

1. Cleaning the pump parts of a handpump has to be done once in month;
2. Cleaning the bottom deposit in a masonary well and the aquatic growth inside has to be done in six months;
3. Disinfection of well water and tubewell parts by chlorine solution specially during rainy season;
4. Bacteriological examination of the water sample can be done from the nearest laboratory.

When there is an epedemic of gastro-intestinal diseases, water must be boiled for drinking purposes. Strong bleaching powder solution (1 mg/l or more) must be put into all open wells and drinking water sources. The source of pollution must be detected by a survey for the control of the same.

Chemical Disinfection

Chlorine is a useful disinfectant for drinking water and is effective against the bacteria commonly associated with water-bone disease. In the usual doses, it is not effective against certain cysts and ova, nor against organisms embeded in solid particles. Chlorine enters almost instantaneously into chemical combination with organic matter in water and in such combined form it is of no use for disinfection. Sufficient chlorine, must therefore be added to satisfy the 'chlorine demand' of water in addition to the amount required for bactericidal action. This combination with other substances may given rise to a chlorine taste, but the presence of such a taste is not a sufficient indication that a free chlorine residual exists.

Chlorine is the easiest to apply in the form of solution. A good stock solution to use in treating water contains about 1 per cent available chlorine.

Bleaching powder or chlorinated lime is a white powder which contains about 30 per cent available chlorine when freshly made. However the strength of the powder rapidly vanished after the can is opened, or in cans that have been stored for a long time.

To make 1 litre of about 1 per cent stock solution, 40 gms of bleaching powder (25% available chlorine) is to be added. Chlorine solutions should be kept in brown or green bottles and stored in dark places.

Well waters can be disinfected by this prepared stock solution. As an example, a dug well of 2 metre diameter having a water depth of 3 metre is to be disinfected. For that, 80 gms bleaching powder of 25 per cent available chlorine is to be added to water of a 21 litre capacity bucket and the supernatent has to be poured into the well. If the water is clear but highly coloured, like very weak tea, or if it has a noticeable sulfur odour, the dosage should be doubled. After dosing, the water should be thoroughly mixed, and left to stand for 20 minutes or longer before being used. Sufficient mixing can be obtained by pouring water from the container in which it is chlorinated into the well.

Sanitation

The programme of sanitation in rural areas consists of:

1. Safe collection and disposal of all solid wastes.
2. Utilisation of all waster matter for improvement of soil fertility.
3. Installation of proper sanitary latrines and urinals.
4. Installation of Bio-gas plants.
5. Installation of better cowsheds and chicken shelters.
6. Installation and use of smokeless chula.

Two Pit Pourflush Latrines

This latrine is found to be very suitable in rural areas because of its low cost as well as it is a completely sanitary latrine. It consists of a squatting pan with steep bottom slope (25 degree - 30 degree) and a tap of 20 mm water seal in a cement concrete-trap floor. After use it is flushed by hand using a small container holding about 1.5 to 2 litres of water. The excreta is carried through a pipe or drain into two honeycomb leach pits, which are used alternately. The liquid in the pits percolates and gases are absorbed by the soil, leaving the solids behind. The pits are used alternately each designed to last for about 3 years. When one if filled it is taken out of use and the excreta is then diverted into the second pit. When the filled pit is left for about 2 years the contents become rich organic humes which is safe for handling.

With simple care pourflush water seal latrine is a very satisfactory and sanitation system-hygienic.

8

Basic Concepts of Watershed Management

Dr. Bimal Chandra Mal

Land is the most precious heritage and physical base of biomass production of life supporting system. The soaring numbers of Human and livestock populations are escalating the demands on biomass production as well as for non-farm uses like urbanisation in a limited area of 329 million hectares of the country. The arable land resource is continuously shrinking. The net sown area per capita was 0.33 hectare in 1950 and going to be 0.15 hectare by 2000 A.D. Wanton misuse is degrading the land resources every day. Almost 175 million hectare covering more than half of the country are threatened by various types of land degradation like saline-alkali soils, water logged area, ravinous and gullied lands, areas under ravages of shifting cultivation, desertification etc. Specific problems of land degradation are due to open cast mining operations, using good productive land for brick kilns, coastal erosion and sea water ingress, excessive erosion and land slides in the crumbling hilly areas. Our forests and grasslands have been over-exploited. Our Ex-Prime Minister Mr. Rajiv Gandhi in his first broadcast to the nation on 5th January, 1985 observed, "Continuing deforestation has brought us face to face with major ecological and socio-economic crisis. This trend must be halted."

---

* Professor, Department of Agricultural and Food Engineering, I.I.T., Kharagpur 721302.

Watershed is defined as a drainage area whose runoff flows through one outlet. The boundary of a watershed can be delineated by using the contour map, various types of land degradation in a watershed, the causes and aims of watershed management are discussed in the following sections:

Types of Land Degradation

1. Saline-alkali soils
2. Water logged areas
3. Ravinous and Bullied area
4. Ravages of shifting, cultivation (Jhumming)
5. Desertification
6. Open cast mining operation
7. Brick kilns in productive lands,
8. Coastal erosion and sea-water intrusion
9. Land slides and excessive erosion in hilly areas
10. Land areas polluted by industrial wastes

Causes of Land Degradation

1. Wrong techniques in the use of cultivated sloping land, up and down cultivation
2. Successive growing of crops that accelerate soil erosion
3. Shifting cultivation
4. Removal of organic matter by burning
5. Use of unbalanced and excessive fertilisers
6. Use of excessive chemicals
7. Faulty method of irrigation
8. Growing of cultivated crops in high sloping areas
9. Destruction of natural protective cover by felling of trees

10. Forest fire

11. Over-grazing of grass land

Aims of Watershed Management

1. To protect, conserve and improve the land for efficient and sustained production
2. To develop without ecological destruction
3. To conserve the soil in situ
4. To reduce the siltation of reservoirs
5. To retain fertility of soil
6. To force water into land profile so as to deliver it slowly to the streams through seepage
7. To build up good ground water potential
8. To reduce flood and sediment damages and safeguard farms, ponds, reservoir etc.
9. To generate rural employment specially during natural calamities

Soil Erosion—Causes and Effects

Principal causes of erosion are:

(i) Destruction of natural protective cover like trees and surface vegetation by indiscriminate felling of trees, forest fire, over grazing of grass lands etc.;

(ii) Wrong techniques in the use of land such as up and down cultivation on the slope, successive growing of crop that accelerate soil erosion, removal of organic matter by burning, unbalanced cropping practices and faulty methods of irrigation.

Damages Caused by Erosion are:

(i) Pollution of stream water

(ii) Siltation in reservoirs

(iii) Loss of nutrients from the soil

(iv) Deterioration into land capability

(v) Fragmentation of land

(vi) Loss of land

(vii) Loss of moisture

The factors that affect erosion are

1. *Climate*

   (i) rainfall

   (ii) temperature

   (iii) wind

2. *Topography*

   (i) degree of slope

   (ii) Length of slope

3. *Vegetation. Major effects are*

   (i) Interception of rainfall

   (ii) Retardation of surface velocity

   (iii) Physical restraint of soil movement by binding effect

   (iv) Increased biological activity to increase the permeability

   (v) Transpiration to reduce soil moisture.

4. *Soil Properties*

   (i) Soil structure

   (ii) Soil texture

   (iii) Organic matter

   (iv) Moisture content

   (v) Compactness

   (vi) Chemical and biological characteristics

Water Erosion is Classified Into

1. Rain-drop erosion
2. Sheet erosion
3. Rill erosion
4. Gully erosion
5. Stream Channel erosion

Measure to Control Erosion and Conserve Moisture

1. *Contour Farming* (1-3% slope)

   Contour farming is the practice of conducting field operations such as ploughing, planting, and cultivating land across the slope rather than up and down hill.

2. *Strip Cropping*

   It is the practice of growing alternate strips of close growing and intertilled crops in the same field. Three types are there:

   i) Contour strip cropping

   ii) Field strip cropping

   iii) Buffer strip cropping

3. *Conservation Tillage*

   It consists of soil manipulation such as ploughing, disking, harrowing and cultivating. Aggregation, surface sealing, infiltration and resistance to wind movement take place.

4. *Mulch Tillage*

   Tillage that leaves the surface of the soil cloddy and mulched with crop residues is an effective accompanying measure with strip cropping to conserve moisture. It breaks the raindrop to dissipate the energy, obstructs runoff and promotes infiltration.

5. *Subsurface Tillage*

Where grain straw and stubble are left from proceeding crop, seedbed preparation by subsurface tillage implement is beneficial. The implements are of 3 types: i) tillers with V-Shaped sweeps; ii) straight blade tillers; and iii) rod-weeder tillers. They loosen and pulverise the soil and cut off the roots of weeds and other plants but do not invert the mulch crop.

6. *Listing and Ridge Planting*

In low rainfall areas in which a large per cent of the rainfall comes in short intense storms, and in regions where gently sloping fields permit the use of contouring alone as an erosion control practice, tillage may be carried out with lister (double M.B. plough). In some cases, a device is attached to create dams at intervals in the channel. This practice is called basin listing or basin tillage. It increases the water conservation in the furrow.

7. *Terracing*

A terrace is an earthen embankment or ridge constructed across a slope to conserve moisture and minimise soil erosion.

Types Based on Functions

(i) Level terrace also known as contour bund follows absolute contour

(ii) Graded terrace used to convey excess water

Based on Construction

(a) Channel type

(b) Ridge type

(i) Narrow base

(ii) Broad base

(c) Bench terrace

Land Capability Classification Based on Slope

Class of land	*% Slope*	*Land Use Pattern*
A. Land suitable for cultivation		
Class-I	0-1	Any crop with proper rotation and green manuring to maintain soil fertility.
Class-II	1-3	Contour farming, contour strip bonding or terracing.
Class-III	3-5	Intensive agronomical measures such as contour cropping, strip and cover cropping, terracing or bunding.
Class-IV	5-8	Contour bunds or terraces and intense agronomic measures. Soil building and soil maintaining crops.
B. Land not suitable for cultivation		
Class-V	8-12	Permanent pasture with controlled grazing.
Class-VI	12-18	Pasture, grass and forestry, with restricted grazing.
Class-VII .	18-25	Forest with restricted felling, contour trenching as conservation measures.
Class-VIII	greater than 25	Forest with complete closure to grazing and felling of trees.

Recomendations for NSS Volunteers

1. Don't remove vegetative cover from hilly areas.
2. In steeply sloping areas, don't grow cultivated grain crops. Grow trees and horticultural crops.
3. For intermediate slopes, grow cover crops or alternate strip crops.
4. In sloping land, plough and plant perpendicular to the slope.
5. Don't make furrow and ridge along the slope.

6. Don't destroy organic matter by burning, make compost.
7. Test the soil and use balanced fertiliser.
8. Don't allow grazing in erosion prone soil. Grow fodder and feed the animals.
9. In low rainfall area, plough the land before rain for moisture conservation.
10. Conserve moisture in situ, don't allow it to go as a waste.

Some Slogans

1. Soil is precious — conserve it.
2. Water is life — conserve it.
3. Soil and Water Conservation is essential — For our survival and food supply
4. Watershed Management is most essential — For control of flood and drought
5. We may build satellites, step into the moon but can not create any soil — Let us conserve the nature's most invaluable gift.

9

Role of Plants in Control of Environmental Pollution

Rajib Lochan Panigrahy

Introduction

Environment is the condition and surroundings that influence our development and that of plants and animals. Environment pollution is the act of contamination of dirty things which are harmful substances to make corrupt the environment/surroundings of species. The harmful substances may be the noise and hazardous gases to air, chemicals and pesticides to water and earth, etc.

According to K.R. Dikshit, 1984 "the space we live in, the air we breath, the food we eat, the water we drink and other resources we need, all these we draw from our environment.

Environment is the sum fall social, economical, biological, physical and chemical factors which constitute the surroundings. It is the combination of all organic and inorganic substances surrounding man.

Ecological Factors Influencing Environment

The living organism may be surrounded and effected by a number of external forces. These forces are environmental or ecological factors as hypsographic, climatic, adaptec and biotic factors.

Elements of Environment

All the biotic and biotic environments can be distinguished in to four segments depending on their physical, chemical and biological characteristic as: (1) atmospheric (air); (2) Hydrosphere (water); (3) lithosphere (soil & land); (4) Biosphere.

Atmospheric (air) pollution

The multi-layered envelop of different gases surrounding the earth which holds up life on earth and saves if form harmful environment of outer space. It extends the height of about 1600kms. from earth. After this, there is no air to exist living organ.

Pollution generally tends to air pollution in recent trend, which directly affects the health hazard of living organ. Environmental pollution is one of the major problems not only in India but also around the globe, which is a looming danger. Pollution is a very common in industrial areas and mega cities due to heavy industrialisation, vehicles and power stations. The anti-pollution machineries are now available in the market to reduce to rate of increasing pollution. Due to expensive cost factors, the power plant, vehicles and industries are reluctant to install, But it is an essential duty to the society.

Growth of industrial activities with scant attention to environmental safeguards, intensive agricultural productivity efforts, high volume of motorised traffic and increasing immigration to urban centres have created ambient air pollution problem in metropolitan growth centres.

Hence, in 1994, Supreme Court of India on the basis of painstaking report by the National Environmental Engineering Research Institute (NEERI), Nagpur recommended plantation of specified kinds of trees with in a radial distance of two to three meters from the Taj Mahal. A tree member's inspection team comprising Advocate Commissioner of the Supreme Court, Mr. Krishna Mahajana and two Central Pollution Control Board environmental engineers Dr. S. Ghose and Dr. Trivedi was sent to Agra in March 1996 to inspect whether 55,000 trees had actually planted around from the airborne pollution and check out if the 25,000 trees shelter belt had been created in the intervening two years since the Supreme Court expressed concern over the ambient air population in the region. The inspecting team submitted its report in April 1996.

Hence, Government made provision to issue permission on building, house planning in cities with having the space for at least a tree near the house. It implies every house have a tree. Don't think it a Government pressure, but a remedy to reduce our environmental (atmospheric) pollution at lest, which should felt in the mind of every people as a responsibility, before deforestation, people should think for forestation. The villagers or streets or cities should reserve an arable land for aforestation having a plantation are for picnic, amusement, take rest in summer days, etc. It can flow a fresh cool air to the environment. So, every people should take care of tress.

In the recent devastating super cyclone, all the trees of the major portions of Orissa has already destroyed. To fill the gap the department of Forest should take innovative effort to fill the gap by plantation. The year old trees cannot found in the recent generation. People afraid of possibility of increase in population and sun stroke in summer days, irregular rainfall so case of flood, draught, etc. in this age trees are ingredients to make alive the living organs on the earth.

Plants as pollution reducing agent

In such a critical phase, plant offers a heavy tool to solve this problems with low cost and availing other socio-economic resources. Some plants are very sensitive to a specific pollution and develop characteristic symptoms. Tobacco plants offer such mechanism to develop flecking symptoms on leaves on exposure of ozone. These plants are helpful in monitoring the population levels precisely. Some plants are resistant to pollutants. In India the Laboratories like NBRI, Lucknow and JNU, New Delhi are working on this problems. They are plants reducing pollutants.

Due to increasing number of factories, automobiles on the road large amount of pollutants are pumped into the air. They include dust, ashes, powers, sulphur dioxide(SO), Proxy Acetyl Nitrate (PAN), nitrogen Oxide, Ozone(O), lead, etc. Vegetation and soil are the sinks for these pollutants. Total 40 per cent of air pollution in India is dust pollution. The gas and power plants, vehicles on road, un-cautious use of several resources like lack of efficient and sufficient knowledge on environment pollution by our uneducated and semi-educated citizen and lack of Government interference to control and reduce it. Trees having compact branching and closely arranged leaves had shown to possess more dust and collecting ability. Alstonia Macrophylla and Dalbergia Bission are the examples of trees reducing dust pollution. The trees having broad leaves of simple elliptical and hairy structure collect a lot of dust from air. The plants are Ficus Bengalensis, Fiscus Infecteria, Tectona Grandis, Shorea Robusta, Polyathea Longifolia and Magnifera Indica which are collecting dusts from air. The trees like tamarind, cassia and neem are collecting dust but poor in ability. Height and size of plant should also be considered as one of the criteria to trap or reduce dust effectively. The plants having shiny and waxy leaves also collect the dust which is easily washed away by rain water. Mass multiplication of the variety trees are essential to cover the polluted area to reduce dust from air and produce more oxygen, consume nitrogen oxide from air which genders the seasonal rainfall and in control.

Some of the Important Plants Tolerant to Gaseous Air

1. Plants tolerant to Sulphur Dioxide—*Polytgia longifolia Terminalia arjuna, Acer plantamides, Thuja orientakis.*

2. Plants tolerant to Ozone—*Zinni elegans, Gladidus sp., Pelargonium gravelones.*

3. Plants tolerant to Oxide of Nitrogen—*Carrissa carandas, Cariacum varigatum, Quercus robur.*

4. Plants tolerant to Lead—*Cassia siamea and Zizyphus mauritiana.*

5. Plants tolerant to PAN—*Acer pantanoides, Chrysanthemum sp.*

6. Plants tolerant to noise pollution in cities—*Ashok, Neem, Coconut and Tamarind.*

7. Plants tolerant to chronic diseases—*Neem, Ashok.*

Thick vegetation helps to absorb or deflect sound waves

Hence, in 1994 the Supreme Court of India on the basis of painstacking report by the National Environmental Engineering Research Institute (NEERI), Nagpur recommended plantation of specified kinds of trees within a radial distance of two to three kms from the Taj Mahal. A three members inspection team comprising Advocate Commissioner of the Supreme Court Mr. Krishna Mahajana and two Central Pollution Control Board Environmental Engineers Dr. S. Ghose, Dr. Trivedi was sent to Agra in March' 1996 to inspect whether 55000 trees had actually planted around Taj Mahal to attempt partially is lolate the heritage monument from the air borne pollution and check out if the 25000 trees shelter belt had been created in the intervening two years since the Supreme Court expressed concern over the ambient air pollution in the region. The inspecting team submitted its report in April 1996.

Hence Government made provisions to issue permission on building house planning in cities with having the space for a tree in the possession of house. It implies every house have a tree. Not through Government pressure, for our polluting atmosphere, plantation is required as the remedy which should felt by every people as a responsibility. Before deforestation, people should think for afforestation. The villages or streets of cities and towns should reserve a space for afforestation. Due to massive depletion of forests people now feel to enjoy plantation areas for picnic, excursion, etc. But, individually every people should take care of trees as ornamental in nature.

In the past few decades many natural calamities like cyclones, floods, tsunami, earth and the constructions of industries, buildings, highways most of trees in the forest, community and road sides has been destroyed. But, through plantation or forestation schemes or by the communities participation even 1 per cent of that has not been required. Which happened atmospheric temperature is going high in every year, irregular and less rain fall and heavy rain fall at a time which cause flood, soil erosion, drought, etc. happens. In this age trees are the ingredients to make alive the living organs in earth.

REFERENCE

1. "Pollution the Looming Danger", *Yojana*, July, 1996.

10

Environment
Prime Concern of Sustainable Rural Development

*Dr. Aditya Kumar Patra**

India is predominantly a 'rural economy' with varied socio-economic structure. Three-fourths of people of India live in rural areas. The rural India is endowed with rich natural resources, huge quantity of common property resources, enriched bio-diverse heritage and a good amount of botanical wealth in the form of varied flora and fauna. However, the rural masses are suffering from abysmal poverty and wide spread unemployment and under-employment. Over the years, rural development has emerged as a strategy designed to improve the economic and social life of a specific group of people—the rural poor. The term 'rural' is broadly referred to the area(s) where people are usually engaged in 'field activities involving production of food, fiber, ores and raw materials. 'Development' is a process of gradual growth or advancement through progressive change. It could be conceptualised as a set of desirable societal objectives that

* Lecture in Economics, Kalinga Mahavidyalay, G. Udayagiri, Kandhamal, Orissa, 762100, E-mail: adityapatra@rediffmail.com

does not decrease over time (Pearce, 1990). These objectives are:

1. Increase in rural per capita income (economic growth).
2. Improvement in distribution of income (equity).
3. Political and economic freedom.
4. Equitable access to resources, education, health care, employment opportunities and justice.

Hence 'Rural Development' may be defined as "A process leading to sustainable improvement (development) in the quality of life of rural people, especially the poor". Rural Development involves provision of jobs and income generating activities while maintaining and protecting the environment of rural areas.

Now a day 'Sustainable Development' has come to the center stage of public policy and academic discussion. The World Commission on Environment and Development (WECD, 1987) defined sustainable development as "development that meets the needs of the present without compromising the ability of future generations to meet their own needs".

Economic development is not an unmixed blessing. On the one hand it brings higher material welfare by increasing the production of goods and services and on the other hand, it degrades the environment badly by the mindless exploitation of natural resources. Natural resources including natural environment perform two important functions in the process of' economic growth: (1) providing inputs to production process; and (2) assimilating the wastes generated in the process of production. As our planet earth is finite, closed and non-expanding one, there is a limit to both these critical functions. Hence, one can not go on producing the goods and services using natural resources in an unlimited manner. There are ecological/natural limits to economic growth and hence it can not be sustained forever. The present environmental crisis is the result of the pursuit of the goal of' maximisation of economic growth using the natural resources in an unscrupulous manner.

In the process of economic development the costs of environmental damage in the form of growing deforestation activities, soil erosion, water pollution, land degradation, loss in biodiversity etc. may out-strip the benefits of having more output of goods and services. The repercussion of environmental degradation have already been visible in the form of green house effects causing acid rain, depletion of ozone layer, rise in global temperature, climatic change, erratic and irregular monsoons, rise in sea level, depleted aquifers etc. may jeopardise the productivity in the long run and put a question mark on the very survival of the rural masses, since the rural economy is dependent on the enrichment of these natural resources.

Now the pertinent question is: how development can co-exist with better environment so that development can be sustained for generations to come.

Sustainable development is possible by keeping the environmental damage within a tolerable limit. It presupposes that in the process of economic growth, we maintain our natural resources and environment intact and use only that much quantity which is regenerated naturally, i.e., we live on the 'flows and keep the `stock' on natural resources and environment intact.

The environment is defined as "the natural world in which people, animals and plants live" to survive and reproduce. The environmental factors that influence the life on earth may be broadly classified as abiotic and biotic elements. The former includes climatic factors related to aerial environment, soil condition, topographic factors etc., while the latter includes the living organisms, Nature has maintained a fine balance among the various entities of environment to enable the plants, animals and human beings to survive, But in the name of economic development this balance is disturbed by the human beings and the net result is environmental crisis. Therefore, there is a need for tailoring the pace and pattern of economic growth to the earth's carrying capacity.

Environmental quality is an integral part of development, Without environmental ethics development is simply meaningless, The fragile environment is undergoing unsustainable levels of stress from growing populations, increasing demand of resources and pollution (air, water and noise) from households, agricultural and industrial sectors causing additional pressure on land and water.

Depletion of our rural environment is mainly on account of the following reasons:

1. Unscrupulous use of common property resources;
2. Excessive use of chemical fertiliser;
3. Imprudent use of water;
4. Poor management of sanitation/garbage;
5. Indiscriminate cutting up of trees for fuel and other purpose leading to deforestation and desertification.

Protection of environment is an essential part of economic development. Without adequate environmental protection development is undermined, Damage to environment affects the welfare of both the present and future generations adversely. It harms human health, reduce economic productivity and leads to the loss of amenities.

To overcome this problem, environmental education plays a key role. Environmental education offers the opportunity to understand the intimate relationship between the quality of environment and well being of the mankind. It also entails practice in decision making and self formulating code of conduct about issues pertaining environmental quality, Environmental education is a process in which individuals gain awareness of their environment and acquire the knowledge and skills to enable them to act individually and collectively to solve future environmental problem. To promote environmental education participation of all the stakeholders viz., Self Help Groups (SHGs), User Groups (UGs), Civil Society Organisations (CSOs), Local Level Institutions (PRIs), Non Governmental Organisations (NGOs), Governmental Organisations (GOs) and mass media is very essential.

Sustainable rural development can promote the development of democratic process at the grass root level by creating awareness on social forestry, biodiversity, soil conservation, population control, protection and restoration of the quality of rural water bodies, rural energy conservation, eco-friendly agricultural activities etc. Hence, sustainable rural development banks on the rational use of scarce resources with the support of people and policies adopting strategic environmental management practices. To achieve the target a multifarious strategy needs to be formulated, to mitigate the rural problems through development process and to protect the environment through community participation.

In the present globalised world, people's participation have become hallmark of rural development. It plays an implicit and inevitable role in deriving the fruits of development from different schemes. The application of participation is rooted in the logic that people should have an opportunity for controlling their own fate as much as possible. Participation has now become virtually synonymous with development projects. Prof. Gunnar Myrdal pointed out that, "the ideal has always been that the plan should come from the people and meet their wishes and need and have their support in thought as well as deed. This is regularly stated in the plans as an important aim of planning and conditions for their success".

A popular vehicle for people's participation and for delivering services in natural resources management is users' groups. Thus water users' association, forest users' groups and watershed committees have been formed to achieve the rational utilistion of natural resources and to meet the equity and efficiency objectives of development.

Let us analyse one such technology known as watershed management practice to achieve the target of sustainable rural development.

The watershed programme was reckoned as a catalyst to bring the second-green revolution in the rain-fed areas. Some people figuratively describe watershed management as a programme that holds the key to solve problems of employment, economy, ecology, export and equity.

The watershed programme in India was initiated more than four decades ago. However, the vigour and seriousness came only during the 1990s. The nature and scope of the watershed programmes were modified over different plan periods and recently tuned to encourage people's participation.

Watershed is not a technology but a concept, which integrates conservation, management and budgeting of rainwater through simple but discreet hydrological units.

Man and environment are interdependent. It is basically the area of survival of the community living within it and drawing its sustenance from it. In resource fragile areas these demands and the claims on the environment are enormous. Any change in the surrounding environment directly affects the people living therein. Watershed development refers to the conservation, regeneration and its judicious use of all the natural resources like land, water, vegetation, animals and human beings within a watershed. Watershed development tries to bring an optimal equilibrium in the eco-space between natural resources, man and animals. In short, watershed development implies rational utilisation of basic resources like land and water for sustainable production.

As watershed development is an inter-sectoral activity, different components have to be taken into account while undertaking watershed development. These include:

- Human Resources Development (Community Development);

- Soil Conservation and Land Management;
- Water Conservation and Management;
- Afforestation;
- Pasture/Fodder Development;
- Agricultural Development;
- Livestock Management;
- Rural Energy Management;
- Farm and Non-Farm Value Addition activities;

All these activities are interdependent and interactive.

Watershed management tries to bring about the best possible balance in the environment on the one side and man on the other. It requires people's participation because those who destroy it would have to want to regenerate and conserve it.

For holistic and sustainable development of rain-fed areas the Government of India now follows an integrated watershed approach. In this approach active involvement and participation of rural communities is ensured which is one of the main component of the farming system approach.

This approach aims at involvement of people through UGs, SHGs, CBOs and also rural institutions in planning along with technological empowerment and capacity building which helps in improvement of farm household employment and income in rural areas on a sustainable basis.

Watershed management becomes a people's movement while Government, donors and NGOs should participate in peoples programme by providing funds, technical advice and social stimuli. People's participation is being recognised by government agencies as an essential component of the success of the watershed management. The implementation of Participatory Watershed Management (PWM) practices has enabled communities to overcome problems and gain more control over their natural resources. People's participation is the core element for the success of PWM process. Under PWM, the local dwellers and neighbouring villagers (termed as stakeholders) participate voluntarily in the entire process.

The basic components of watershed management as enumerated above are considered strategic in successful PWM designated under a three-fold stratification of:

1. Water Management Policy
2. Biomass Management Policy
3. Determining long-term sustainability consideration.

Water Management Policy includes Rain Water Harvesting, Ground Water Recharge. Maintenance of Water Bodies, Preventing Water Pollution and Economic Use of Water.

Biomass management includes Eco-preservation, Biomass Regeneration, Forest Management and Conservation. Plant Protection and Social Forestry, Development of Animal Husbandry Sector, Income and Employment Generation activities along with co-ordination of Health and Sanitation.

Determination of Sustainable consideration in PWM is qualitative in nature. So, there is every chance of de-motivating on the part of the stakeholders. However, quantitative generalisation regarding the sustainability considerations are tried to be formulated on the basis of the following premises:

Socio-Cultural Indicators

1. Decision-making power of the community
2. Empowerment of women
3. Formation of User Groups/Self Help Groups
4. Harmonious social life
5. Improvement in quality of life such as migration, education, health and hygiene.

Economic Indicators

1. Increase in income and employment
2. Availability of food and food security
3. Improvement in standard of living
4. Off-farm income to families
5. Improvement in rural economy
6, Availability of credit and saving
7. Availability of water, fuel and fodder

Environmental/Ecological Indicators

1. Increase in productive potential of resource base
2. Management of Common Property Resources (CPRs)
3. Improvement in the bio-diversity of the region

The number of components in different watershed management programme may vary according to the objective set and strategies made for the accomplishment of the goal.

PWM seems useful but not so simple in practice. It involves two serious problems:

1. It is a physically difficult task;
2. The elite groups in the community tries to block such noble moves when it feels that it would be a looser in the process either directly (diversion of water) or indirectly (benefit flows from a project go to others).

Considering such non-conducive environment the following steps are suggested to make PWM more effective:

1. Low cost, simple and vegetative consideration technology based on indigenous technical know-how should be promoted with watershed village people's initiative and with minimum external interference/support;
2. Ensure women's participation in all levels of PWM;

3. Expansion of general awareness and literacy for motivating the stakeholders to reap the multi-dimensional benefits of PWM.

The steps for sustainable rural development should be pro-poor and environmental friendly so that it will lead to an ever growing production base without disturbing the natural ecological balance. Hence. We conclude with the words of Anil Agrawal, a renowned environmentalist, that *"Rural Development should become holistic, at the same time, village eco-system should be protected"*.

REFERENCES

Agrawal Anil (1985). Ecological Destruction and the Emerging Patterns of Poverty and People's Protests in Rural India. *Indian Social Institute,* New Delhi.

Khanka, S.S. & Anjan Bhuyan (2006), Participatory Watershed Management, *Kurukshetra,* Vol. 54. No. 9. pp. 12-16,

Mukundan, N., (2004), Environment, People's Participation and Sustainable Rural Development: An Overview, *Kurukshetra,* Vol. 52, No. 8, pp. 1-6.

Nagasree, K. & K.V. Suhramanyam (2006), Farming Systems Approach for Rural Development, *Kurukshetra,* Vol. 54, No. 8, pp. 22-25.

Pearce David, Edward Barbier & Anil Markadya (1990), Sustainable Development: Economics and Environment in the Third World, *Earthscan Publishers Ltd.*, London.

Seth, S.L., (2000), Watershed Management in India, *Kurukshetra,* Vol. 48, No. 9, pp. 2-7.

Technical Aspects in Watershed Development, *Watershed Organisation Trust,* Ahmednagar.

WCED (1987), Our Common Future. World Commission on Environment and Development. *Oxford University Press.* New Delhi.

11

Environment Audit and Innovative Reprimand *Need of the Day*

*Nibedita Padhy**

The concept of environment is much wider, namely, environment for an unborn child is his/her mother's womb and for a newly born child, it is the lap of his/her mother or father. Generally environment is equated with nature wherein the physical components of the planet Earth, viz., water, air, land and the inter relationship which exists among water, land, air, plants, micro-organisms, human beings and other living creatures. In fact, people implore for the inclusion of only 'air-land-water-plant' in the concept of environment, thus excluding human society from the sphere of environment. But, in the correct context, man forms an inseparable part of the environment.

The word environment has been derived from a French word 'environ' means surroundings. Environment refers to the sum total of biotic and abiotic components surrounding and influencing the man at a given point in space and time. The environment is a source of energy and materials which mankind transforms into goods and services to meet his social, cultural, economic and political needs. As regards to

* **Lecturer, PG, Department of Biosciences, CPS, Mohuda, Berhampur, Orissa.**

the biotic environment, the human beings can be grouped under physical, economic, social and political in their activities and characteristics. The physical man consuming all the basic elements of physical environment transforms it into a sink of all types of wastes and pollutants. The economic man always uses the resources of environment with his innovative skill, knowledge, wisdom and technologies to fulfill his need. When the activities of the economic man goes beyond the limit it causes detrimental to the environment. The social man is the pressure group, who constitutes social institutions and NGOs draws the attention of the political man. In the other hand the political man is the law maker, policy maker and frames rules, regulations, laws for social welfare and the protection of his own existence.

The Article 21 of Constitution of India gives the fundamental right to every person 'the right to life and environment' i.e. 'Enjoyment of life—including(the right to live) with human dignity encompasses within its ambit, the protection and preservation of environment, ecological balance free from pollution of air and water, sanitation, without which life cannot be enjoyed... Hygienic environment is an integral facet of right to healthy life and ... there is a constitutional imperative on the State governments and the municipalities, not only ensure and safeguard proper environment but also an imperative duty to take adequate measures to promote, protect and improve both the man-made and the natural environment.' [Supreme Court of India in Virendrer Gaur vs. State of Haryana 1995 (2) SCC 577].

With the growing population on the globe and having limited land and water gifted by the nature causes heavy pressure on the natural resources including land and water. Hence we must desist from over exploitation of the resources of the environment as the earth provides enough stuff to satisfy every man's need, but not every man's greed. Time has come when we are facing challenge for saving the humanity from extinction because our activities have been constricting the environment and depleting the biosphere.

Environmental pollution and its effect on global warming is a major threat to living system on earth. Global warming is the rise in the average temperature to a level which influences the existence of life. Global warming will not only lead to loss of flora, fauna and human population on earth but also cause a major reshuffle to cropping pattern of the world along with devastative natural calamities. It is the outcome of variety of factors, both natural and anthropogenic like fossil-fuel burning, deforestation, rapid industrialisation, urbanisation, soil pollution, water pollution, vehicular emission etc.

In the present chapter a brief study has been made about the environmental pollution in Orissa on account of multifaceted problems.

The state of Orissa, extending over an area of 1,55,842 sq. km. lies on the east coast of India bordering Bay of Bengal between 17° 48′-23° 34′N latitude and 81° 24′-87° 29′E longitude. It is bounded with Jharkhand and West Bengal on the north, Chattisgarh on the west, Andhra Pradesh on the south and Bay of Bengal on the east. It ranks 11th with respect to total population spreading in 105 towns and 51,124 villages. The total area of the state is 1,55,707 sq. km and the coastline stretches over a distance of 482 km. covering nine districts from a total of thirty revenue districts.

God has gifted us a suitable, polluted free natural environment. It helps every species on earth for their easy survival. But the insatiate greed of man is threatening the healthy and diverse biological order and spoils the natural environment of organisms creating a menace upon the ecology. The environment of Orissa has been changed radically in the last decades. Given below are a few global and local problems which have been mainly responsible for squalor of environment in Orissa.

Deforestation

Forests play an important role in the economy of the State. But the pressure of human population and livestock on forests for fuel wood, small timber and grazing is much more and beyond the carrying capacity of existing forests. According to the Forest Survey of India's present forest cover in the State is 31 per cent as against 33 per cent of the total geographical area mandated by the National Forest Policy, 1988. Thus there is an imperative need to accelerate efforts for a forestation and regeneration of the degraded forests for achieving the mandated 33 per cent forest cover.

Orissa has a recorded forest area of about 58,136.23 sq. km. with the following classifications:

Table 11.1

Types of forests	*Area in Sq. Km*
Reserve Forests	26,329.12
Demarcated Protected Forest	11,686.44
Undemarcated Protected Forest	3,838.78
Unclassed forest	20.55
Village Forests, Khesra Forest and Others	16,261.34
Total	58,136.00

In spite of above, extensive and unabated deforestation is going on at an alarming rate causing unprecedented threat over the ecosystem and the state is facing problems like:

1. Land erosion and land slides;
2. Irregularities in the pattern of rain fall;
3. Frequent natural calamities like flood, drought and cyclone;
4. Disappearance of many important and rare wild animals and birds;
5. Migration of some specific animals and birds from their natural habitats to other places in search of food and shelter;
6. Loss of Biodiversity;

7. Increase in the concentration of green house gases resulting global warming.

It is conspicuous to note that since last six years construction of Grand Trunk Road is going on converting NH 5 into a four way GTR. It has compelled the road constructors during the process of expansion of NH 5 to fell a number of old banyan and other trees which were providing oxygen and cool breath to the passers by. But no steps have been taken yet either by the Department of forest, Government of Orissa, NGOs or by civil society activists to plant trees along side the newly constructed Grand Trunk Road (GTR) to protect the environment.

E-waste

With the development of science and technology the use of television, refrigerators, washing machines, telephones, radios, music systems, computers and mobile phones have increased several folds among the people of Orissa. All such electronic goods are replaced by their users at every three to five years posing a sever challenge of disposal when the waste is generated from them and later on these wastes are exposed to the environment. According to a latest study Orissa generates about two thousand tons of e-waste annually and it will definitely increase in pro-rata in accordance with the growth in the consumption of electronic products. According to I.T. research firm IDC India, the total number of PCs in India has already crossed 22 million units at the end of March 2007, suggesting one PC of approximately every 50 Indians due to advancing technology, lower prices and global demand in the part of technology aware population. In pursuance of the consumption scenario of PCs it has been assumed that there will be more than one billion PCs in use by the end of 2008. Similarly, the use of mobile phones and also the use of other domestic electrical implements will also increase due to its availability at a affordable price. Therefore, strict steps should be taken from right now for the recycling of e-wastes. It is high time for the Orissa Pollution Control Board (OPCB) to access the gravity of endangers which people of Orissa will confront tomorrow on account of the hazardous materials like lead, silver, cadmium, mercury, selenium, chromium etc. are released from e-wastes. The people of the country are using electrical and electronic goods for more than four decades but it is surprising that there is no proper regulatory guide-lines for handling of wastes released from them. The manufacturers, local authorities like Panchayats, N.A.C, Corporations and PCBs should take up this challenge, failing which the users of e-goods have no choice except to put the obsolete or broken electronic item in the garbage.

Excessive Mining

Orissa has rich natural resources like aluminium, chromite, bauxite, iron, coal etc. Orissa holds the prime place in chromite reserves and production accounting for more than 95 per cent of the country's chromite wealth. In Orissa chromite deposits are mostly located in Sukinda valley of Jajpur district and Boula-Nuasahi belt of Bhadrak district. To assess the pollution potential of chromite mines in the Sukinda Valley and Coal mines in Talcher of Orissa, detailed survey and environmental monitoring were conducted by the Central Pollution Control Board, Eastern Zone office. It was observed that one tone of chromite mining generates about ten tones of over burden on water bodies. According to their report the chromium released from chromite mines was being discharged into adjacent water bodies in its hexavalent form. The over burden was being dumped near the queries as a result soluble metals may percolate to the ground water causing ground water pollution. If this process will continue time may come people may not get pure drinking water. The total rate of production of chromite in Orissa is increasing year after year as reflected below in table 11.2. But it is a matter of regret that there are no adequate treatment facilities for this problem. Mining of these minerals from the rocks also causes soil erosion making the mine area prone to natural calamities like land slides, earth quakes, floods, cyclones etc.

Table 11.2: **Production of Chromite in Orissa in MT**

Districts	*2000-01*	*2002-03*	*2004-05*
Dhenkanal	43288	17241	12261
Jajpur	1737442	2976081	3035201
Keonjhar	61705	70972	75924
Orissa	1842435	3064294	3123386

Festivities and Pollution

Festivals like Ganesh Chaturthi, Deepavali, Dashera, Laxmi Puja etc. contribute pollution to air, water and noise in villages, towns and cities. People celebrate all kinds of victories right from winning in an election to observe immersion of deities, and in the functions including marriage reception, thread ceremony and Bhagawati Jagaran by bursting crackers causing air pollution. Similarly loud speakers, fire crackers and sound come from vehicular horns at the loudest decibel level cause noise pollution. It causes both physiological and psychological problems. Besides hearing loss and stomach disorder it can result in irreversible changes in the nervous system and affect the heart patients seriously. The Supreme Court banned the use of loudspeakers, vehicular horns and fire-crackers between 10 pm. and 6 am. Responding to a Public Interest Litigation (PIL) petition, a two-member division bench asked the government to consider restricting loud speakers even during the day. But no one is abiding by the rules. Countries like the US, Australia, Japan and China have legislations to control noise pollution, which yet India has to adopt. While on the issue of noise pollution it is worth recalling that the Supreme Court has literally put a cap on "noise" our crackers be capable of emit. It is 125 decibels on an average; the ruling has had a useful effect. The immersion of the idol in local tanks, ponds, rivers and mostly in oceans pollute the water and cause a threat to the water bodies, animals and human beings.

The Central Pollution Control Board (CPCB) has identified about 24 areas as critically polluted. Some of them are given below:

Table 11.3

State	*Critically Polluted Areas*
Delhi	Najafgarh Drain
Uttar Pradesh	Singrauli
Bihar	Dhanbad
Maharastra	Chembur & Tarapur
West Bengal	Durgapur & Howrah
Assam	Digboi
Punjab	Govindgarh
Himachal Pradesh	Parwanoo
Rajasthan	Pali & Jodhpur
Gujarat	Ankaleshwar & Vapi
Madhya Pradesh	Korba
Orissa	Talcher
Tamilnadu	North Arcot
Karnataka	Bhadravati
Andhra Pradesh	Visakhapatnam and Patancheru-Bollarm belt near Hyderabad

The above named polluted places deserve focussed attention of the policy makers, administrators and the concerned pollution creating agencies to take steps for immediate improvement of the quality of environment prevailing in the aforesaid locality.

Indoor pollution

It is estimated that 75 per cent to 80 per cent of water pollution takes place because of domestic sewage. A recent report of the World Health Organisation (WHO) asserts that indoor pollutant released is one thousand times more likely to reach people's lungs than a pollutant released outdoors. It has been estimated that about ten lakh women and children die each year from indoor air pollution in our country. It works out to almost 28 per cent of such deaths world wide.

The only plausible reason for this phenomena is the fact that rural India mostly uses 'dirty household fuels' like animal dung, crop residues and wood, which are at 'the bottom of the energy ladder'. Electricity, which is the cleanest fuel, is beyond their reach and so is natural gas or liquefied petroleum gas (LPG).

Cooking on an open coal stove exposes women to risks of lung cancer, according to the WHO. A National Family Health Survey report says use of biomass fuels is contributing to pulmonary tuberculosis as well. It is a matter of surprise that indoor pollution yet has not drawn much attention.

Industrial Pollution

The major congregation of industries are found in the districts of Cuttack, Puri, Sambalpur, Sundargarh, Balasore, Ganjam and Koraput. The district-wise Industrial Scenario is furnished below:

Table 11.4: Medium and Large-Scale Industries

Factory	*Location*	*District*
Steel plant	Rourkela	Sundergarh
Fertiliser plant	1. Rourkela	Sundergarh
	2. Talcher	Angul
Mig factory	Sunabeda	Koraput
Machine Building	Kansbahal	Sundergarh
Cement factory	Rajgangpur	Sundergarh
	Baragarh	Baragarh
Ferro manganese	Joda	Keonjhar
	Rayagada	Rayagada
Ferro vanadium plant	Rajgangpur	Mayurbhanja
Ferro chrome	Jajpur road	Jajpur

(*Contd…*)

Factory	*Location*	*District*
Ferro silicon	Theruvali	Rayagada
Tube factory	Choudwar	Cuttack
Orient paper mills	Brajarajnagar	Sambalpur
Ballar paper mills	Choudwar	Cuttack
J.K. paper mills	Rayagada	Rayagada
Emami paper mills	Motiganj	Balasore
Textile (OTM)	Choudwar	Cuttack
Konark paper mills	Keonjhar	Keonjhar
Sewa paper mills	Jeypore	Koraput
Sugar factory	Aska	Ganjam
	Nayagarh	Nayagarh
	Badamba	Cuttack
	Baragarh	Baragarh
	Rayagada	Rayagada
Auto India	Rourkela	Sundergarh
Glass factory	Hirakud	Sambalpur
Aluminium factory	Hirakud	Sambalpur
Cable factory	Hirakud	Sambalpur
	Bhubaneswar	Khurda
Re-rolling mill	Hirakud	Sambalpur
Tile factory	Choudwar	Cuttack
Kalinga iron works	Badabil	Keonjhar
Salt factory	Humma	Ganjam
Custic soda factory	Supandi	Ganjam
Graphaite crucible unit	Titlagarh	Bolangir
Rusmat	Bhograi	Balasore
Hydroelectric project	Balimela	Malkangiri
	Kolab	Karaput
	Machhakund	Karaput
	Indravati	Nabarangpur
	Hirakuda	Sambalpur
Refractory	Raj gangpur	Sundergarh
	Belphar	Sambalpur
	Baranga	Cuttack
Spinning mills	Baragarh	Baragarh
	Jharsuguda	Jharsuguda
	Choudwar	Cuttack
	Rairangpur	Mayurbhanja
	Aska	Ganjam
	Rahama	Kendrapara

(Contd....)

Factory	*Location*	*District*
K.K. Synthetics	Angul	Angul
Rare earth	Chhatrapur	Ganjam
Orissa chemicals	Talcher	Angul
Cane and bamboo handicraft	Bidyadherpur	Cuttack
Betelnut handicraft	Aska	Ganjam
Orissa lamps	Khurda	Khurda
Orissa plastics	Mancheswar	Khurda
Konark TV Factory	Bhubaneswar	Khurda
	Jagatpur	Cuttack
Railway coach repairing	Mancheswar	Khurda
Jute mills	Dhandamal	Cuttack
	Rupsa	Mayurbhanja
Jayashree chemicals	Ganjam	Ganjam
Birla tyres	Balasore	Balasore
NALCO	Damanajodi	Koraput
	Angul	Angul
Cycle factory	Cuttack	Cuttack
Matches factory	Talcher	Angul
Oil Orissa	Khurda	Khurda

This scenario will change very soon, adding more number of industries in the state as a good quantity of MoUs has been signed by the Government of Orissa with various Multinational Companies (MNCs).

The effluents from these industries contain many inorganic and organic pollutants like lead, mercury, zinc, copper, cadmium, chromium etc. that are toxic to living organisms. Industrial effluents may also contain some radioactive substances which cause serious health hazards. Water Initiatives Orissa (WIO) dealing with water issues, has published alarming findings in the cover story of WIO's bi-monthly Oriya language Newsletter, *'Panira Dagara'* (the Water Messenger) in its latest Jan-Feb 2006 issue that in the last 13 years, from 1991-92 to 2004-05, severely degraded land in the state has increased by 136 per cent, barren land has increased by 69 per cent and land converted to non-agricultural uses has increased by 34 per cent. This is about

7 per cent of Orissa's total geographical area. By 2004-05, as high as 17.5 per cent of Orissa has turned barren or unsuitable for agriculture. The way mineral and water guzzling heavy industries are being pushed, the rate in which forest cover is thinning in the state, the rapidity with which the climate is showing changes and the rate at which degradation is increasing, the state may turn a desert even faster. In mining and industrial districts like Raygada and Jharsuguda agriculture lands are already shrinking very fast. In Raygada district un-productive lands is nearly 174 per cent higher than cultivable land. Such lands amount as high as two/third of Jharsuguda's total cultivable area. "The way lands rapidly becoming barren and degraded, desertification is anything but a reality now," observes WIO. Conservative estimates have put that if the proposed steel plants starts functioning they will alone emit 392 million tonnes of Carbon Dioxide by 2010. Further, those industrial units, at the present count, will require at least 527 millions of clean water and release those as pollutants. Washington based Institute of Policy Studies (Daphne Wysham "Destroying Orissa, Fuelling Climate Change," Institute for Policy Studies (USA), April 1997) has cautioned that by 2010 Orissa will emit 7 to 10 per cent of Global Green House Gas emission.

The following laws have been framed to meet the aforesaid challenges but how far the laws are successful in checking the industrial pollution will speak if environment audit will be introduced.

1. Water (Prevention and Control of Pollution) Act, 1974.
2. Air (Prevention and Control of Pollution) Act, 1981.
3. Environment (Protection) Act, 1986.
4. Central and State Pollution Control Boards.
5. Ganga Action Plan, for restoring the quality of river by Central Ganga Authority.
6. National River Action Plan.
7. Environmental Impact Assessment by Ministry of Environment and Forests, Government of India.

8. Promotion of Effluent Treatment Plants.
9. The Water Mission.
10. Department of Non-conventional Energy Sources, for new renewable non-polluting sources of energy as solar, wind, tidal, biomass-based sources, biogas.
11. Recycling of wastes.

Marine Pollution

Orissa has 30 districts out of which 9 are coastal districts. Along with others the habitats of coastal districts are mainly responsible for marine pollution. The types of marine pollution seen along Orissa Coast can be classified according to their nature, source, physical state etc. The most widely known are as follows:

(a) ***Sewage pollution:*** Sewage is considered as one of the major pollutants of ocean water as the sewage is rich with microbes, organic components like carbohydrates, proteins, fat, urea etc. and inorganic salts such as nitrates, phosphates from detergent and soaps with sodium, potassium, calcium, chloride etc. from different substances. These components favour the growth of water-borne pathogenic micro flora and algae causing Eutrophication. These microbes and organic components present in water also cause some serious diseases like cholera, typhoid, dysentery, jaundice, poliomyelitis, hepatitis etc. in human beings and hamper the growth of acquitting organisms.

(b) ***Heavy metal pollution:*** With regard to metal pollution in the ocean let us take the example of sinking of ships and marine vessels for a number of proximity causes. Similarly aeroplanes falling into the oceans due to unforeseen accidents like birds strikes, engine disorder, shortage of fuel, failure of radar etc. It is seen from a statistics maintained in the office of Comptroller and Auditor General (CAG) that there were 135 air

accidents in 2002 out of which 24 were due to bird strike. In the year 2003 the rate of accidents increases to 165 out of which 30 were due to bird strike. In 2004 there were 142 air accidents with 27 birds strike. If the probable ratio of planes falling into the ocean is simply 5 per cent the metal pollution inside the ocean causes a great disorder marine biodiversity.

(c) ***Oil pollution:*** Oil released from industries, ships and vessels into the ocean seriously cause harm to the marine animals and plants.

(d) ***Pesticide pollution:*** Pesticides from crop lands and industries changes the water quality of rivers, seas and oceans and cause toxic effects to the living organisms.

(e) ***Industrial pollution:*** Industries are the source of different types of organic and inorganic pollutants in the marine water. The detailed figures of the industrial pollutants discharged by different industries are given below:

Table 11.5

Pollutants	*Annual estimated discharge (in metric tons)*
1. Petroleum and hydrocarbons	3,405,000
2. Industrial and airborne hydrocarbons	15,000,000
3. Lead compounds	350,000
4. Mercury compounds	100,000
5. Benzene hexachloride	50,000
6. D.D.T	25,000
7. Polychlorinated biphenols	25,000

Marine pollution has seriously affected the exploitable living resources, recreational and commercial uses of ocean and the overall integrity of marine ecosystem. Marine pollution put adverse effect on climate. From a study it is known that due to climatic change the visit of migratory birds has been decreasing year by year in the Nalban area of Chilika Lake. For the same cause different species of turtles and fishes are dying in the seas of Gopalpur and Paradeep. Hence, protection of sea from pollution becomes very essential in ocean resources management. The effective planning for controlling and combating marine pollution requires knowledge about the magnitude of pollution, the routes of entry, physicochemical characteristics of pollutants, and their impact on living organisms.

Places of Tourist Interest and Pollution

Tourism is one of the largest industries in the world. Orissa is not an exception to it. Orissa has many places of tourist interest. Tourists during their visit to tourist places throw polythene bags, disposable packets, straws, plastics, ghootka packets and so on. Pollutants disposed by tourists in three different tourist places of Orissa published in the statistical bulletin are given in Table 11.6, 11.7 and 11.8. *(See on pages 150-155)*

Population Explosion

Population explosion is the root cause of all environmental problems. The population on the globe is the total number of human beings living on the planet Earth at a particular point of time. The world population from 1950 to 2010 presented by US Census Bureau, International Data Base, August 2006 version. (See the Graph on page 156).

The growth rate of world population as shown in the graph is increasing in every census, though different regions have different rate of population growth. Over population adversely affects the environment like:

1. Non-availability of fresh water for drinking, as well as sewage treatment and effluent discharge;
2. Depletion of natural resources, especially fossil fuels;
3. Increased level of air pollution, water pollution, soil contamination and noise pollution;

Table 11.6: Pollutants Disposed by Tourists at Rambha near Chillika Lagoon during November 2004 to April 2005

Month	*Date*	*Polythene and Plastic*		*Straw*		*Disposable Paper*		*Others*	
NOV	07.11.04	65	Avg.80	32	Avg. 32	38	Avg. 43	32	Avg. 31
	14.11.04	75		32		40		30	
	21.11.04	85		35		46		32	
	28.11.04	95		38		48		33	
DEC	09.12.04	77	Avg. 87	34	Avg. 39	44	Avg. 49	30	Avg. 34
	16.12.04	79		37		44		34	
	23.12.04	97		42		52		36	
	30.12.04	99		45		56		40	
JAN	03.01.05	103	Avg. 126	42	Avg. 51	55	Avg. 48	38	Avg. 43
	10.01.05	127		47		51		42	
	17.01.05	129		55		45		44	
	24.01.05	147		62		41		48	

Table 11.6: (Contd....)

Month	*Date*	*Polythene and Plastic*		*Straw*		*Disposable Paper*		*Others*	
FEB	01.02.05	101		52		36		42	
	08.02.05	95	Avg. 84	46	Avg. 46	35	Avg. 33	40	Avg. 36
	15.02.05	76		44		33		33	
	22.02.05	67		42		31		30	
MAR	01.03.05	65		32		31		25	
	17.03.05	57	Avg. 54	30	Avg. 26	29	Avg. 26	22	Avg. 21
	24.03.05	50		24		23		20	
	31.03.05	45		21		22		18	
APR	08.04.05	43		20		19		18	
	15.04.05	36	Avg. 32	13	Avg. 12	17	Avg. 16	15	Avg. 13
	22.04.05	29		10		15		13	
	29.04.05	22		8		13		9	

Table 11.7: Pollutants Disposed by Tourists at Barkul near Chillika Lagoon during November 2004 to April 2005

Month	*Date*	*Polythene and Plastic*		*Straw*		*Disposable Paper*		*Others*	
NOV	03.11.04	60		25		37		36	
	10.11.04	70	Avg.75	37	Avg. 39	32	Avg. 39	29	Avg. 33
	17.11.04	80		45		43		32	
	24.11.04	90		51		46		35	
DEC	05.12.04	78		32		40		27	
	12.12.04	85	Avg. 91	36	Avg. 37	44-	Avg. 46	35	Avg. 37
	19.12.04	98		39		49		42	
	26.12.04	103		42		54		45	
JAN	06.01.05	107		39		56		33	
	13.01.05	122	Avg. 129	42	Avg. 45	61	Avg. 64	30	Avg. 28
	20.01.05	136		48		69		26	
	27.01.05	151		53		73		20	

Table 11.7: (Contd....)

Month	Date	Polythene and Plastic		Straw		Disposable Paper		Others	
FEB	04.02.05	110		42		51		26	
	11.02.05	94	Avg. 86	52	Avg. 55	48	Avg. 44	20	Avg. 19
	18.02.05	76		61		41		19	
	25.02.05	64		65		36		11	
MAR	07.03.05	71		24		30		18	
	17.03.05	60	Avg. 58	21	Avg. 19	29	Avg. 24	12	Avg. 12
	24.03.05	55		19		21		14	
	31.03.05	48		15		18		9	
APR	02.04.05	42		18		28		17	
	09.04.05	35	Avg. 34	14	Avg. 12	23	Avg. 20	14	Avg. 13
	16.04.05	39		9		19		13	
	23.04.05	22		7		11			

Table 11.8: Pollutants Disposed by Tourists at Satapada Near Water-spread Area during November 2004 to April 2005

Month	*Date*	*Polythene and Plastic*		*Straw*		*Disposable Paper*		*Others*	
NOV	04.11.04	65		23		35		33	
	11.11.04	78	Avg.78	31	Avg. 31	37	Avg. 40	31	Avg. 32
	18.11.04	82		34		43		33	
	25.11.04	89		37		45		34	
DEC	07.12.04	81		33		41		31	
	14.12.04	83	Avg. 91	35	Avg. 38	44	Avg. 46	34	Avg. 35
	21.12.04	97		41		49		37	
	28.12.04	103		44		53		41	
JAN	05.01.05	107		41		55		39	
	12.01.05	131	Avg. 130	46	Avg. 50	61	Avg. 62	41	Avg. 36
	19.01.05	133		54		63		34	
	26.01.05	151		61		71		31	

Table 11.7: (Contd....)

Month	*Date*	*Polythene and Plastic*		*Straw*		*Disposable Paper*		*Others*	
FEB	02.02.05	105		51		53		43	
	09.02.05	99	Avg. 88	54	Avg. 45	49	Avg. 46	41	Avg. 37
	16.02.05	80		43		43		34	
	23.02.05	71		41		32		31	
MAR	03.03.05	69		31		34		26	
	10.03.05	61	Avg. 58	29	Avg. 26	33	Avg. 31	23	Avg. 22
	17.03.05	54		23		31		21	
	24.03.05	49		21		29		19	
APR	04.04.05	47		19		29		19	
	11.04.05	40	Avg. 26	12	Avg. 11	27	Avg. 24	16	Avg. 14
	18.04.05	33		9		21		14	
	25.04.05	22		7		11		10	
						20			

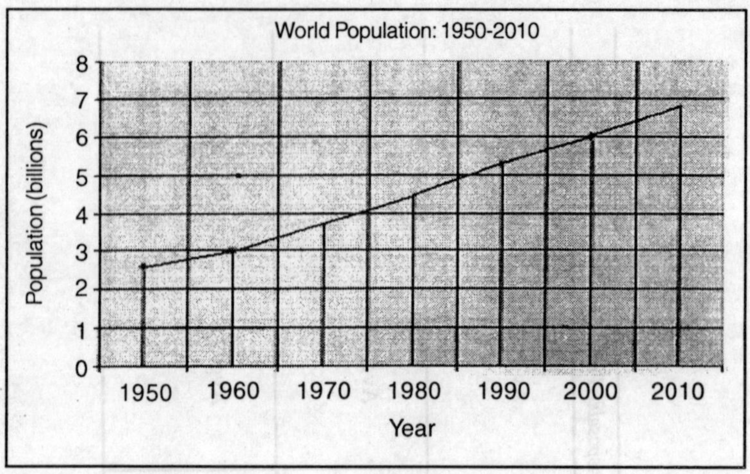

Source: U.S. Census Bureau, International Data Base August 2006 version.

4. Deforestation and loss of ecosystems that sustain global atmospheric oxygen and carbon dioxide balance; as about eight million hectares of forest are lost each year;
5. Changes in atmospheric composition and consequent global warming;
6. Irreversible loss of arable land and increases in desertification;
7. Mass species extinctions. From reduced habitat in tropical forests due to slash-and-burn techniques that sometimes are practised by shifting cultivators, especially in counties with rapidly expanding rural populations; present extinction rates may be as high as 140,000 species lost per year;
8. Increased incidence of hemorrhagic fevers, HIV and other infectious diseases from crowding, disturbance of ecological systems;
9. Starvation, malnutrition or poor diet with ill health and dietdeficiency diseases (e.g. rickets);
10. Poverty coupled with inflation in some regions and a resulting low level of capital formation;

11. Low birth weight due to the inability of mothers to get enough resource depletion, discharge of raw sewage and solid waste disposal;
12. Unhygienic living conditions for many based upon water resource depletion, discharge of raw sewage and solid waste disposal;
13. High rate of unemployment in urban areas (leading to social problems);
14. Elevated crime rate due to drug cartels and increased theft by people stealing resources to survive;
15. Conflict over scarce resources and crowding, leading to increased levels of warfare;
16. Over-utilisation of infrastructure, such as mass transit, highways, and public health systems.

Population explosion is a burning problem of our country. India occupies second place in the world in population whereas it occupies 7 place in terms of area. India has 2.4 per cent of world's land with 16 per cent of world's population. The nature has given us air, land and all other resources with limited means. Unless a positive and effective check has been made on the growth of population, no doubt India will overtake China and we will face a number of problems like food, unemployment, health and what not? So, government and public as a whole should have to think seriously and remedial measures have to be taken to solve these problems. Orissa is also embraced with over population and the growth rate of population in Orissa is also increasing at an alarming rate. The growth of population of Orissa is shown in Table 11.9.

Shifting Cultivation

In Orissa about 80 per cent of population depends on agriculture. Mostly the tribal population in Orissa practices shifting cultivation. Shifting cultivation is carried out by burning large areas of forest in the hilly areas causing in calculable damages like:

Table 11.9: The Growth of Population of Orissa

Year	*Population*	*Decadal Variation In Orissa /(India)*	*Progressive growth rate over 1901 in Orissa /India*
1901	10,302,917	—	—
1911	11,378,875	10.44 (5.75)	10.44 (5.75)
1921	11,158,586	–1.94 (-0.31)	8.31 (5.42)
1931	12,491,056	11.94 (11.00)	21.24 (17.02)
1941	13,767,988	10.22 (14.22)	33.63 (33.67)
1951	14,645,946	6.38 (13.31)	42.15 (51.47)
1961	17,548,846	19.82 (21.51)	70.33 (84.25)
1971	21,944,615	25.05 (24.80)	112.99 (129.94)
1981	26.370.271	20.17 (24.66)	155.95 (186.64)
1991	31,659,736	20.06 (23.86)	207.29 (255.03)
2001	36,706,970	15.94 (21.34)	256.28 (330.80)

1. Exponential loss of soil fertility;
2. Damage the ecology and environment of that region;
3. Declination of natural forest;
4. Disappearance of native species due to fragmentation of habitat;
5. Encourages invasion of other plants.

To overcome the aforesaid problems a new technology called Sustainable Agriculture Land Technology (SALT) has been introduced in Orissa which suggest mixed cropping pattern. But the villagers face some problems in carrying out this technology like lack of irrigation, paucity of capital for investment and lack of infrastructural and technical support for processing and marketing of their products. The tribal people also do not have the legal rights over the land they cultivate. All these problems can be avoided if the Scheduled Tribes Recognition of Forest Right Bill, 2005 will become an Act.

Soil Pollution

With the increasing rate of population, the rate of soil pollution or land pollution is increasing. Now the natural soil is threatened by different activities of man. Industrial solid wastes, mining wastes, agricultural chemicals and domestic wastes are the main pollutants of soil. The non degradable wastes like plastic, metals, glass, ceramics and poly-bags cause more threats to our soil than degradable wastes. So these wastes should not be thrown away to the soil. It needs proper collection and recycling process. Every individual and local authority have important role in collecting these solid wastes for proper recycling process. Laws have been established to ban the use of poly bags but no one strictly obeys these laws. If man will sleep over his devastative activities then time will come when our mother land will be unable to support the biodiversity. As warned by Water Initiatives Orissa (WIO) the eastern Indian state of Orissa will turn to a mass of barren and desert like lands in another 150 years,. This is an alarming finding. Visualising the fact the whole world had observed last year as the year of deserts and desertification with the theme, "Let's stop dry lands from turning deserts". Desertification is a process of productivity loss of lands. When severe, it leads to permanent damages. WIO has reached this conclusion using State data, Orissa Economic Survey; rainfall and temperature data from Climatologically Data of Orissa that, many parts of Orissa, specifically the western and southern uplands, have developed symptoms of desertification; they have further degraded from drought prone to desert prone areas.

Urban Housing

The demand for urban housing is increasing day by day for several demographic factors, namely, size of population, changes in fertility, mortality, migration and new family formation by unplanned additions to and cutting of trees and spoiling the cultivated lands. Housing problem is an untold story in metros. Slums found in urban scenario, are one of the worst types of environmental degradation and have become concomitant to urbanisation and industrialisation. National building organisation observation reveals that about 18.75 per cent of India's urban population lives in slums. The slum dwellers live in an environment with inadequate living space, sewage facility, water supply causing human health hazard with a steady deterioration of surrounding regions. Urbanisation has a strong adverse impact on environmental quality. Indeed Urban environment is considered as the opposite of the natural environment. People in urban areas suffer from breathing problem due to air pollution. Chemical pollutants are also found in the urban environment, which cause ill health and premature death. This challenge is met in different ways as follows:

1. Human Settlement Schemes to deal with the problem of urban slums. Steps are being taken by the National Commission on Urbanisation, 1988 and Housing and Urban Development Corporation (HUDCO), 1970;

2. Policies on housing, occupational health, slum improvement etc.

But unless and until the local authorities does not introduce environment audit the environmental pollution in urban areas can't be checked.

Vehicular Emissions

On account of the availability of mopeds, scooters, bikes and cars at an affordable price and easy finance given by public and private sector banks people are inclined to purchase more number of vehicles. The increased number in use of vehicles emits more pollutants. Vehicle exhausts release a mixture of chemicals including carbon monoxide, sulphur dioxide, nitrogen oxides and hydrocarbons. Some gasoline also contains lead which appears in the exhaust fumes. This can have adverse effects on brain development in children. Four wheelers are running in the state without registration under Motor Vehicle Act using kerosene instead of diesel and contribute pollution in the environment. Such activities should be strictly banned.

Effects of Environmental Pollution

Global warming and Global dimming/global cooling are the outcome of the above natural and anthropogenic factors. Naturally the green house gases like carbon dioxide, water vapour trap heat that radiates from the earth's surface and act as a thermal blanket surrounding the earth. Otherwise the earth's average global temperature would have been -18°C (-4°F). But due to the use of refrigerators, air conditioners, paints, sprays, fertilisers and hydroelectric reservoirs, some other green house gases like cholorofloro carbons (CFC), nitrogen monoxide (NO_x), methane (CH_4) and ozone (O_3) intensely increased the temperature of earth. Table 11.10 shows that how these gases are responsible for global warming.

Table 11.10

Green House Gases	*Atmospheric Concentration*	*Contribution in Green House Effect*	*Heat Absorbing Capacity More than CO_2*	*Source*
CO_2	≈ 350ppm	57%		Fossil fuel burning by automobiles, power plants and factories
CH_4	≈ 1.7 ppm	12%	25 times	Bacerial decomposition of flooded peat and forest biomass, Hydroelectric reservoirs
CFC	≈ 0.32 ppm	25%	15,000 times	Refrigerators, air conditioners, paints and spray manufacturing industries
NO_x	≈ 0.32 ppm	6%	230 times	Nitrogan fertiliser, vehicular emission
0_3	≈ 0.02 to 0.1 ppm		2000 times	Developmental activities
Water vapour	≈			

We have already increased the rate of CO_2 by 25 per cent. The concentration of CO_2 is increasing at a rate of about 1.8 mg/m^3 per year. If the concentration of CO_2 will increase in this rate then it is expected that by the year 2030 the rate of CO_2 will be doubled. The day is not far off when India, the sixth largest emitter of greenhouse gases, would be compelled to accept some limits on its emissions. This possible development has to be viewed in the light of the country's growing energy needs. According to the "Hydrocarbon Vision 2020" prepared by the Planning Commission, even in 2020 the country will be dependent on fossil fuels such as coal and oil for a major share of its energy requirements. The implications of this prognosis for India's contribution to global warming clear. The first will be enhanced and the second will deteriorate. The NEP is silent on these developments. Information gathered from weather department indicates that while the global mean temperature rose by 0.5 degree Celsius in the last 50 years that of Orissa rose by 1 degree Celsius. Weather is becoming alarmingly extreme. In a decade, average of highest recorded temperatures has increased by 4.4 to 6.6 degree Celsius and the average of lowest recorded temperatures has further decreased by 3 to 5.1 degree Celsius in various parts of Orissa.

Effects of Global Warming

1. Global warming will not only lead to loss of flora, fauna and human population on earth but also cause a major reshuffle to cropping pattern of the world.
2. Arise of 5 degrees would raise the sea level by five meters within a few decades, threatening all the densely populated coastal cities of India. Arise in sea level of 50-100 cm caused by ocean warming would flood low-lying lands in Bangladesh, West Bengal and Orissa.
3. Due to green house effect and early snow melts in mountains there may occur more hurricanes, cyclones and more floods during monsoon.

4. Climate change could lead to the extinction of many animals and migratory birds. Global warming has already changed the migration routes of some birds.
5. It is damaging the breeding colonies of sea birds and sea animals.
6. Warmer seas strongly affect the sex ratio of hatchlings and it can change some turtle species into entirely female.
7. Growing water scarcity could destroy the wetland areas.

Global Dimming

Like Global warming another major problem of environmental pollution is Global dimming. It means the sun is dimming gradually and the earth surface is experiencing less light now-a-days in compared to recent past. The scientists recorded at least 10 per cent drop in sunshine reaching the earth surface from 1950s to 1990s. This event was first brought to public in 1989 by Atsumu Ohmura, a researcher of Swiss Federal Institute of Technology. This was conformed and the term Global Dimming was 1st coined by Dr. Gerald Stan hill of Israel's Agricultural Research Organisation in 1992.

The main cause of Global dimming is fossil fuel burning and in some extent the use of high-flying planes. From these sources green house gases along with some particulate matters such as soot, ash and vapour are released to the atmosphere. More particulate matters produce large number of water droplets. These larger water droplets in the atmosphere reflect sunlight back into space and cause global dimming.

Recently scientists observed that over the next century there is five degrees increase in global warming. But if the global dimming effect could be eliminated the temperature rise would be ten degrees instead of 5 degrees and the earth will get hotter.

Thus now the main problem lies in the fact that the global dimming problem or global warming problem can not be solved as a separate environmental problem. The demand of the time is to deal with both these problems together and steps should be taken at the earliest. The NEP is silent on the role of urban local bodies in environmental improvement. These institutions are poorly endowed with finances and lack expertise in managing the local environmental problems. For example, the Municipal Waste Management Rules, 2000 have yet to be enforced in most urban areas due to financial and other constraints. Capacity building among the staff and devolution of adequate powers to them by the State Legislatures should find a place in the NEP. The 73rd Amendment to the Constitution provides the backing for initiatives in this regard.

Finally, the NEP has not recognised adequately the potential of the After Governments in improving environment quality. After all, most of the subjects that would fall under the definition of the term 'environment' are within the law—making powers of the State Legislatures.

Conservative estimates have put that if the proposed steel plants starts functioning they will alone emit 392 million tonnes of Carbon Dioxide by 2010. Further, those industrial units, at the present count, will require at least 527 millions of clean water and release those as pollutants. Washington based Institute of Policy Studies (Daphne Wysham "Destroying Orissa, Fuelling Climate Change," Institute for Policy Studies (USA), April 1997) has cautioned that by 2010 Orissa will emit 7 to 10 per cent of Global Green House Gas emission.

In the aforesaid circumstances the Planning Commission should take due care of environment in their approach in the 11th Five Year Plan. But it is a matter of regret that the Planning Commission has not seriously considered the environmental problems, the states and the country is confronting today, as expressed by Sri Mohan Dharia, the former Deputy Chairman, Planning Commission. There are a number of environmental Acts in force in India.

They are: *Water Prevention and Control of Pollution Act, 1974*—This is an Act to provide for the prevention and control of water pollution and the maintaining or restoring of wholesome of water, for the establishment, with a view to carrying out the purposes aforesaid, of Boards for the prevention and control of water pollution, for conferring on the assigning to such Boards powers and functions relating thereto and for matters connected wherewith. After 1988 amendment, a Board may close a defaulting industrial plant or withdraw its supply of power or water by any administrative order; the penalties are more stringent and a citizen's initiative provision bolsters the enforcement machinery.

The Water (Prevention and Control of Pollution) Cess Act 1977—The Act creates economic incentives for pollution control through a differential tax structure (with higher rates applicable to defaulting units) and requires local authorities and certain designated industries to pay a cess for water consumption. These revenues are used to implement the Water Act 1974.

The Air Prevention and Control of Pollution Act, 1981—This is an Act to provide for the prevention, control and abatement of air pollution, for conferring on and assigning to such Boards powers and functions relation thereto and for matters connected therewith. The 1987 amendment empowered the Board to shut down the defaulting industrial plant or may stop its supply of power or Water, Air Pollution, also included noise pollution.

The Environment Protection Act, 1986—This is an Act to provide for the protection and improvement of environment and matters connected there with:

- Environment (Protection) Rules 1986;
- Hazardous Wastes (Management and Handling) Rules 1989;
- Rules for the Manufacture, Storage and Import of Hazardous Chemicals, 1989;

- Rules for the Manufacture, use , Import and Storage of hazardous Micro-Organisms/Genetically Engineered Organisms of Cells, 1989;
- The Coastal Regulation Zone Notification 1991;
- Environmental Impact Assessment Regulations 1994;
- Bio-Medical wastes (Management and Handling) Rules, 1998;
- Noise Pollution (Regulation and Control) Rules, 2000;
- MSW (Management and Handling) Rules, 2000.

The Indian Forest Act, 1927—This is an Act witch includes the provisions for hunting restrictions in reserved or protected forests and authorised establishments of sanctuaries. The Act provides both for the public and private forests and facilitates the extraction of timber for profit.

The Forest (Conservation) Act, 1980—This is an Act to provide for the conservation of forests and for matters connected therewith or ancillary or incidental thereto.

The Wildlife (Protection) Act, 1972—Provides for the statutory framework for protecting wild animals, plants and their habitants. It specifies endangered species, which are protected regardless of location, and all species protected in designated areas called sanctuaries and national parks.

The Mines and Minerals (Regulation and Development) Act 1957—This statute promotes the prospecting of minerals and the development of mines. In 1986 amendment, provisions are introduced requiring greater environmental sensitivity whilst conducting mining operations.

The Insecticide Act 1968 and Insecticide Rule 1971 along with Tiwari Committee Report mandates for effective monitoring pesticide residues in the environment.

The Atomic Energy Act 1962 and the Radiation Protection Rules 1971—The regulation of nuclear energy and radioactive substances in India is governed by the Atomic Energy Act 1962, and the Radiation Protection Rules 1971. The Central Government is in mandate to prevent radiation hazards, guarantee public safety and the safety of workers handling radioactive substances, and ensure the disposal of radioactive wastes.

But, the law is provided for those who wake and not for those who slumber and sleep. We have passed a good deal of time by creating awareness among the people on important environmental issues through paper and electronic media, NGOs etc. which are in-vain. It is, therefore, to make every person accountable; introduction of Environment Audit and Innovative Reprimand is the need of the day. Introduction of innovative reprimand has many advantages. First, reprimands like community service, Slum area cleaning will act as a more effective restraint for others than fines and imprisonment. The idea of introduction of innovative reprimand is to have deterrence in public views. When the offenders serves in public places, onlookers get the right message that if they indulge in polluting the environment that could happen to them too. We need to encourage such nobel reprimands with the motive not to punish but to reform. Hence, every local authority must appoint environment scientists as environment auditors. They should make environment audit regularly moving around the arena of their jurisdiction. They may also receive appeals from the inhabitants of their locality who are suffering in environmental pollution due to the carelessness of other persons. Here, Let us define the term person. The term Person includes an individual, a firm, a company, an association of persons whether incorporated or not, local authority and every artificial juridical person. It is high time that the activities of every person contributing to environment pollution should be audited and imposed with innovative reprimands, otherwise the jeopardised condition of the flora and fauna will increase in an alarming rate. It is therefore, along with introduction of environment audit, sufficient budgetary provision should be made in the plan periods for the disposal, removal and treatment of various pollutants of air, water and land through bio-remediation.

Article 51-A of Constitution of India included by 1976 Amendment describes the fundamental duties of every citizen of India, which reads as follows:

- to value and preserve the right heritage of our composite culture;
- to protect and improve the natural environment including forests, lakes, rivers and wild life, and to have compassion for living creature;
- to strive towards excellence in all spheres of individual and collective activity so that the nation constantly rises to higher levels of endeavour and achievements.

In the present juncture a few suggestions are given below which every citizen of India must abide by to live in a dignified manner and let leave others in the same status too:

- Save Oil—Save Energy
- Harvest Rain Water
- Conserve Water
- Recycle Paper/Water
- Plant and Nurture Trees
- Segregate Your Household Garbage into Biodegradable and Non-biodegradable Waste
- Don't Litter Indiscriminately
- Use Jute or Cloth Bags Instead of Plastic Bags
- Use Compact Fluorescent Light (CFL)
- Install Solar Heating/Lighting
- Adopt Green Building Technologies

- Environmental Awareness
- Greater Participation
- More Human Approach
- Use of Bio-gas
- Organ Farming

REFERENCES

1. Dr. A.N. Rai, Environmental Education, *Goyal Brothers Prakashan*, New Delhi.
2. Malladi Rama Rao, *Kurukshetra*, December, 2006.
3. Pollution in Orissa—*A Status Report*.
4. Ranjan K. Panda—Impending Ecological Anarchy: Orissa Will Turn Into a Barren Land.
5. *Statistical Bulletin 2004*, Department of Tourism and Culture, Government of Orissa and Personal Communication.
6. *The ENVIS Newsletter*—June, 5, 2006.
7. *The New Indian Express*—May, 8, 2006.

12

Interlinking of Rivers in India
Some Reflections

Dr. B. Eswar Rao Patnaik♣
Rajib Lochan Panigrahy♠
Santosh Patnaik♥

Irrigation is the foundation on which development of a country rests. Ever since the creation of "Task Force" for linking rivers project, by the N.D.A. Government, the topic linking rivers in India has occupied the center stage in development economics. The N.D.A. government has contemplated on interlinking thirty six rivers of the country for transferring water to various arid zones of the nation. Thus, stage is set for lively debate on the economics of river linking projects in India. This chapter endeavours to study the economics of linking of rivers in India. This chapter is based on secondary data. Five sections are designed. Section I deals with introduction. Section II takes us to a trip on Ganga-Cauvery link as proposed by Dr. K.L. Rao. Section III dilates on benefit cost principle and the blessing side of dams. Section IV deals with the demerits of dams leading to conclusion in Section V.

♣ Senior Reader and HOD (Economics), S.B.R.G. Womens College, Berhampur, Orissa.

♠ Assistant Prof. in Computer Science and Management, Ambedkar College of Management and Technology Berhampur, Orissa.

♥ Er. TC.S. Kolkata.

SECTION - I

Introduction

It is possible to state that, river linking project is not altogether novel to water resources personnel, and scientific community of the country. The eminent Pilot Dastur has added fresh materials to the body of linking river projects by developing the garland canal concept. Arthur Cotton's insistence on linking rivers for accelerating the pace of transport development of the country deserves mention. As far as in 1960 Dr. K.L. Rao has published his thought provoking work "Garlands of Rivers Project" which has mooted on linking the rivers of the country. Pioneering intellectual work of noted thinker, Annand Swaroop Arya has enriched the literature on linking rivers.

The process of linking rivers Raine-Rone, Raine Danube has enabled the European nations to climb high on socio-economic ladder. The secret of the story of amazing agrarian and transport development of Chinese economy is linking rivers project. It is admitted on all hands that, the Indians backed the perception that, conceives rivers as national assets, irrespective of the regions in which they flow. The sharing of waters of rivers Krisna and Yamuna has been the bone of contention between States, Andhra Pradesh and Karnatka. Global action may be guided on conservation principle which. Stresses that, water may be abundant in one place but it is scarce in others. Water is a fundamental right and this yellow gold of the 21st century must be provided equitably to all people and not on the basis of ability to pay.

Commenting on the N.D.A Government's decision on interlining rivers, Dr. N.K. Rath (1993) has observed that, very little information is available in published form to interested students about project linking of rivers in India. Analysts argue that, the programme for linking of rivers does not form an integral part of the Tenth five year plan in India. A. Vaidyanathan (1993) questions the necessity, desirability and feasibility of linking of peninsular rivers, which involves an outlay of Rs. 5,60,000 crore. The answer to the question how the government can mobilise colossal resources to implement the project escapes our logic. Recently, Suresh Prabhu, the Chairman of the Task Force for implementing river projects has candidly admitted that, even preliminary feasible studies have not yet been made. So, it is imperative that, an indepth study of environmental impact, displacement of people and submergence of villages consequent upon linking rivers is made.

SECTION - II

Ganga-Cauvery Link

The principal objective underlying Rao's work "garland of rivers projects" was to establish a link between rivers, Ganga and Cauvery. A sensitive approach to water concerns in designing and implementing water development programmes is a must for reserving water scarcity in regions like Telengana and Rayalaseema districts of Andhra Pradesh. Rao has focussed on fourteen major rivers of the country namely Sindh, Ganga, Brahmaputra, Sabarmati, Narmada, Tapi, Subarnarekha, Brahmani, Mahi, Mahanadi, Godavari, Krisna and Cauvery.

Rivers have been classified into three categories by Rao. The first category rivers included ayacut covering a distance of 20,000 sq.km. and above. Ayacut covering a distance of 2,000 sq.km. to 20,000 sq.km. included second category rivers. The third category rivers included/ayacut covering a distance of less than 2,000 km. central to Rao's proposal is the construction of a canal that links rivers Ganga and Cauvery covering a distance of 2400 km.

The link was to take off near Patna, Pass through the basins of rivers Sone, Tapi, Narmada, Godavari, Ponmar and join the Cauvery upstream of the anicut. The general impression was that the link would withdraw 60,000 cusecs of water from the flood flows of Ganga for about 150 days

Table 12.1: Increase in Availability and its Disposition Arising from the Interlinking of Pennisular Rivers as Proposed by Central Water Commission (bcm)

Basin	*Volume Transferred out of Basin*	*Enroute Utilisation*	*Losses*	*Volume Transferred out of Basin*	*Net Increased in Availability*
Mahanadi	11.2				-11.2
Enroute		-3.9	.80		
Godavari	25.5			6.5	19
Enroute		7.5	.68		
Krishna	16.5			17.3	+08
Enroute		5	.75	(-)	
Ponmar	8.6			11.5	+2.9
Enroute		3.2	.39		
Cauvery	2.3			5.0	+27
Below Cauvery		2	1		
Total	64.1	21.6	2.86	42.3	+6.4

in a year and presumably it may entail a lift of a sizeable part of the water above 450 meters. It appears that, the project may result in transfer of water from Brahmaputra and Ganga westwards to Southern U.P., Harayana, Rajasthan and probably southwards to the peninsular components. The peninsular components embrace several links, of which the premier ones would be those connecting the Mahanadi, the Godavari, the Krishna and Cauvery.

The National Water Development Board which was directed to enquire into the feasibility of the innovation proposal of Rao has recommended steps for linking rivers above Vindhya mountains in Southern region. The initiative of linking rivers below Vindhya mountains was not advocated by the Water Board, owing to its financial implications.

SECTION - III

Benefit-Cost Princple and Case for Linking Rivers

The social benefits of a project undertaken includes, pecuniary and real benefits. Real benefits accrued to the consumers by a project are reflected in addition to community welfare. Pecuniary benefits denote increase in earnings consequent upon changes in relative prices of factors and products *(see table 12.2)*. While tangible benefits are evaluated in terms of market prices, intangible benefits and costs are evaluated in the form of effects on the health and living standard of people, using shadow prices.

Table 12.2: Illustration of Project and Costs in Evaluation of Efficiency Criteria of Public Expenditure to an Irrigation Project

		Benefits	*Costs*
Real			
Direct	Tangible	Increased farm Output	Cost of Pipes
	Intangible	Beautification of Area	Loss of Wilderness
Indirect	Tangible	Reduced Soil Erosion	Diversion of Water
	Intangible	Preservation of Rural Society	Destruction of Wildlife
Pecuniary	Relative improvement in position of farm equipment industry		

Analysts agree that, the ratio of Net Present Value of the Project (NPVP) or the benefits cost ratio provides a measure of economic feasibility of the project. In case of a single project, it is choosen in its B/C ratio is one or greater than one. In case of several competing projects, ranking of projects are in descending order, on basis of B/C or B-C/C.

However, future uncertainties need considerations, while using internal rate of return, so that, present value = future value X discounting factors. The social marginal productivity of a project is evaluated by deducting social opportunity cost of displaced persons from marginal product.

SMP= MP-SOCIAL OPPORTUNITY COST

Case for River linking Projects.

It is imperative that, dams and reservoirs are built in the economy to accelerate the process of inter linking rivers.

The insistence of planners and policy makers on dams derives, its strength from that, agriculture till today, provides sustenance to 60 per cent of people of the country. The lacklustre performance of plan exercises in irrigation frontiers is evident from the fact that, as high as 60 per cent total cultivated land does not have irrigation facilities. So, since independence, the extension of irrigation facilities in farmer's filed in rural and urban areas has been a important component of national endeavour. One bright spot in the body of river linking project is the possible improvement in the per acre yield of crops and agricultural progress, which may attend the creation and utilisation of irrigation in an economy.

Data analysis of the estimate of control Water Commission (Table 12.1) places the gross volume of water which includes both ground water and surface water transferred out of five basins , Mahanadi, Godavari, Krisna. Periyar and Cauvery at 65 billion cubic meters. The figure works out to 62 bcm. Providing for enroute losses to the extent of 2.7 bcm. The gross value realised in the transferee basin is 39 bcm., much of which is transferred from one basin to another. Allowing for this, net availability in the five basins increase by about 24 bcm. It is clear, therefore that, the projected expansion of irrigated area at the national level may be related to the increasing requirement of food and fibre production for a growing population.

Power Generation

Power sector development in a country is the hallmark of modern civilisation. As an economy develops, their will be enormous demand for use of electricity, from all quarters, namely agriculture, industry and transport sectors. So, expansion of capacity generation is indispensable to cope with the surge in demand for the use of power, one feather to the cap of mammoth multipurpose projects is the creation of capacity in power generation. A case of study of power sector development in Orissa reveals that , the five Hydro-Power projects in state sector, Hirakudh, Balimela, Machkund, Rengali and Upperkolab will create an installed capacity of 37.5 megawatts, 360 M.W. 34.5 M.W., 250.0 M.W and 320.0 M.W, respectively.

Employment Generation

It is plausible to believe that, planning is an instrument for betterment of people's living conditions. The generation of large scale employment opportunities in an economy will fight the battle of abysmal poverty.

The line of reasoning of advocates of plans is that, the filed for employment base of the economy stands to the broadened, thanks to the policy thrust on dam construction. The objective is realised by construction of the roads, creation of transport links and growth of swarm like economy activities, like opening of sales and purchase outlets, pisciculture and set up secondary and ancillary industries, which is a logical corollary of construction of dams.

Abatement/Mitigation of Droughts

Human memory is alive to the nightmarish droughts of 1987-88 of country level, which has acted as barrier in the path of agricultural progress of the economy and inflicted untold miseries on the people of the country. It is well recognised that, severe droughts are recurring phenomenon in Kalahandi District of Orissa and Rayalaseema District of Andhra Pradesh. Even now, there has not been any progress in the direction of sagacious and meaningful drought management in the country. It has been well reasoned by protagonists of mega projects that, the creation of mini watersheds and water-harvesting structures in an economy may be an answer to the challenges of droughts. The provisions of water to water scarce regions may facilitate the pursuit of rabi crop or kharrif crop in an area and to that extent, it may lighten the burden of drought on the economy. Micro-watershed management involves total resources conservation including soil and water, forest, horticulture and crop husbandry.

The case for dams rests on safety, they provide against floods. This is achieved partly by storing surplus water of rivers and partly by diverting water to water-stressed regions of the country. History bears testimony to the fact that, the Cuttack District of Orissa was saved from the wrath of floods, times without number by Hirakud dam.

Case Against Dams

1. *Higher Costs of the Projects:* The official estimate does not specify the rate of interest of the inflation in estimating the total cost. Dr. N.K. Rath (2003) assumed a 7 per cent rate of interest to be charged on the capital during the construction period. The total cost of the Himalayan, the Hydro electric and peninsular will come to Rs. 11,478.73 instead of Rs. 5,60,000/crore as originally estimated.

2. *Unworkable Nature of the Scheme:* A. Vaidyanathan (2003) reasons that, the project may comprise 35/30 odd projects, each of which, may have to undergo estimation and clearance. We may not be sure in advance that, all of them will be cleared and if so, it is unclear how the total mega project can be made.

3. The river-linking project is not the answer to the major needs arid zones and drought prone areas of the country. The line of reasoning presented by A. Vaidyanathan runs thus:

The project does not purpose to take water from rivers to riverless areas but merely to Link Rivers. By linking river

A to river B by gravity we can succeed only in transferring some water to river B, which will then flow down hills again by gravity. This may make a little water available to areas already served but how this takes water to India's uplands and dry lands needs analysis.

Submergence of Villages

In the present century, environmental degradation has emerged as a global concern for human survival. The mammoth irrigation project will submerge villages, after installation and tribal population living in village areas would lose their land. Fernandos (1994) has estimated that, during 1951-90, 213 lakh people where displaced because of development projects. The sad tune in the song of Narmada project is submergence of 37 villages comprising 1000 families and 3.5 lakh hectares of India's best deciduous forest comprising teak and bamboo plants will-be submerged in the process. It seems that, the cost benefit analysis of projects made by experts, takes into account only tangible economic values, like timber and minor forest produce and discounts invaluable ecological values, like water replenishment, wildlife shelter, siltation and climatic stabilisation. Reference may be made of the recent promise made by Chief Minister of Orissa, that "Records of Rights" of oustees of Huirakudh Dam would be provided soon. The policy of government should be as far as possible to give land in exchange of land and that too well in advance of submergence.

It would not be out of contest to mention here that the task for implementing Linking Rivers project does not have independent professional secretariat. The organisation is manned by professional from irrigation bureaucracy who have not shown much interest in developmental works.

It is an established fact the river valley projects are not a positive proposition in India. Usually a project is Sanetioned by Planning Commission only, when the cost benefit ratio is 1:1.5. Numerous studies at country level have shown that, the cost benefit ratio is 1:0.46. The suggestion, therefore, is that big dams are not, as beneficial as claimed by proponents of big dams.

Conclusion

The foregoing analysis has presented the balance sheet of river valley projects. Major dams have brighter as well darker sides. In lively debates questions have been raised by scholars of all hue regarding the viability of high dams. The alternative suggested is construction of a number of smaller dams instead of one dam or smaller structure. If mega projects become millions of small projects full economic costs are likely to be recovered. Minor irrigation works and small dams with minimal impacts of water logging and salinity deserves attention. The success that dams can achieve in mitigating the wrath of floods, largely depends on careful maintenance of embankments for avoiding breaches in embankments, and preparation of inundation maps for every breach of embankment. The perception that rivers belong to the nation has to be understood by the people of the country. The daunting task of public authorities is to provide for an effective rehabilitation policy for project affected persons supply land for land, ensure employment of the oustees in suitable lines, fix compensation for loss of land at just price and accelerate the development of social and economic infrastructure around the places, where oustees are rehabilitated. All this depends on the collective wisdom of policy makers public authorities and irrigation experts. The various aspects of the project for interlinking rivers, conceptual technical, economic and environmental need judicious and meticulous review by independent experts. The Brunt Land Commission counsels us "The world in which poverty is endemic will always be prone to a ecological and other catastrophes". As the Vedic seers state succinctly *"Aapsu amrutam"*

Of late the accent of the economy has been on globalisation and liberalisation, which have by passed social welfare considerations for sake of profit maximisation. It seens that, the proposals of major irrigation projects are underway in the economy. Analysts policy makers and socially oriented persons of the country should raise the scale of endeavour to consider the balance sheet of mammoth irrigation projects and evolve alternatives to costly and environmentally less friendly dams.

This chapter is an enlarged version of the paper presented at the 87th Annual Conference in Indian Economic Associations Held at Benaras University in 2004.

REFERENCES

1. A. Vaidhyanathan: Interlinking Pennisular Rivers: A Critique, *E.P.W.*, July. 5, 2003.
2. Ramaswamy Aiyar, Letters to Editor, *E.P.W.*, July 1, 2003.
3. N.K. Rath "Linking Rivers": Some Eleminatray Airthmatic, *E.P.W.*, July 2003.
4. E.E. Nadu "Nedu Jala Sandhanam Sudura Swapnana", *Himabindu*, 10 Aug. 2002.
5. Eswar Rao Patnaik "Large Dams, Sustainable Development and Problems of Displacement". S.N. Tripathy (ed.) Economic Development and Problems of Displacement, 2003, Anmol, Publishers.
6. Editor, *Indian Express*, July 18, 2004.
7. Interview with B. Krishna Rao Patnaik. "Executive Engineer" Khariar.
8. S. Anand Krishnan "Is the Water Bomb Tickling for India". *India Today*, May, 2003.
9. S.N. Behera, "Displacement and Development. A Cost Benefit Analysis, Published in *Economic Development and Problems of Displacement*, (ed.) S.N. Tripathy.

Index